SCIENCE IN THE REAL WORLD

A simplified story of how technology using chemistry and physics is used in the real world of industry

Alan McGown

This is an IndieMosh book

brought to you by MoshPit Publishing
an imprint of Mosher's Business Support Pty Ltd

PO BOX 147
Hazelbrook NSW 2779

www.indiemosh.com.au

First published 2015 © Alan McGown

Cataloguing-in-Publication entry is available from the National Library of Australia: http://catalogue.nla.gov.au/

Title: Science in the real world: a simplified story of how technology using chemistry and physics is used in the real world of industry

Author: McGown, Alan

ISBNs: 978-1-925353-44-0 (paperback)

                     978-1-925353-45-7 (ebook – epub)

                     978-1-925353-46-4 (ebook – mobi)

Contents

Alan McGown

Acknowledgements

There are many people who helped me in getting this book into being and without whom I could not have produced it:

Bruce Neville, Principal of a high school near Sydney for helping to define the use for such a book.

Denise Bailey, Science Coordinator of a high school near Sydney, for reviewing the book and critical comments on its applicability to the syllabus being taught.

Ashley Ahern for her brilliant skill in designing several suggested cover pages; obviously of which I could choose only one.

Maria Angel for her assistance in locating the many diagrams and pictures which are most necessary to explain the technology in such a book as this.

Jennifer Mosher, my understanding publisher, for putting up with the interminable alterations and my odd suggestions about layout and formatting.

And lastly to my long suffering wife Margaret, for putting up with me during the many hours taken in writing.

Diagrams and photographs used in this book are kindly provided under license by Wikimedia Commons or have been created specifically for the purpose.

Introduction

Whether you already have an interest in Science or are not sure what Science is, then this book is designed to show you in a simplified way, how science is used.

It is the story of a few industries where I worked, which were using the latest, most up to date, scientific methods or technologies, and these continue to be used today.

At any stage in this book, you can skip ahead if any subject is of interest and just use the basics chapter to help you understand something such as chemical formulae, which needs some explanation.

As you will see in this book, the processes used in industries described here, always have a basis in science. For example, paper making is still the same basic process it has always been, since it began hundreds of years ago, but the machinery is getting more sophisticated, larger machines make paper faster and wider than before, and more instrumentation allows better control of the process. Chemical additives allow more recycled paper to be used and higher quality paper and cardboard to be produced and for waste products to be minimised.

Another example in more recent times, is the way technology is used in exploration and mining. Core drilling from the surface provides the data needed by the computer. The cores from the boreholes are tested for chemical and physical properties, this data is fed into the computer, and the computer then provides a picture of the physical location of the seam of mineral, and allows planning of the best method of mining or stripping of waste overburden, When mining is due to start, Global Positioning System (GPS) is used by surveyors to mark out the exact location of the deposit. In some mines the vehicles have no onboard driver, but are driven from the office using a computer connection.

Each description of a process is a very simplified version of what actually happens, and is just an overview, so that the reader can appreciate the basic processes without the complication of the numerous other steps. For anyone who is interested in any particular topic, there is a huge amount of information on available on the internet, and I provide at the start of each chapter, some key words to help the reader who wishes to research further on any topic.

I am always amazed at the amount of technology used in industry, most of which the general population is quite unaware. I am seeking to show those students of science based study that the more you learn, the more you realise how little you really know about what is happening in the wide world.

I find the chemistry and physics behind the technology interesting, and I hope I can make it interesting for the reader. Usually in physics there are lots of formulae and calculations. I have left all of that out, to just give simple descriptions of what goes on.

As I am an industrial chemist, I see things from mostly that point of view and there are some chemical formulae provided, which I have tried to keep brief, but some things need an explanation in a little detail to understand what is going on, and sometimes the chemistry needs to be shown to give a proper understanding. A quick look at the basic chemical reaction of a process gives the simplest explanation of what is taking place.

I started work as a laboratory assistant and later graduated to chief chemist or laboratory manager in various industries, and still later in various marketing and management positions. In the end I went back to working at the bench and set up and managed several laboratories for testing coal, before going into retirement.

When I was starting out in my first job, I only knew that I might find it interesting working in a chemical laboratory. I had no idea or plan of where I might be employed in the future. By studying Science, Maths and English at school and Chemistry at Technical College, I was always in a good position to apply for employment in many different industries. As I moved from job to job I was able to use my previous employment experience to show a new employer how much I had learned and how that might be useful to the new employer.

In secondary school, the subject selections of **Science** with **Mathematics and English** will give the student the most options to be available with regard to further education and employment. Leave out any one of

Alan McGown

those three, and you are making a severe limitation to your choices for your future employment.

Chapter 1
A few basics to help understanding

Key words

element, atom, electron, electron shell, nucleus, proton, neutron, molecule, compound, covalent bond, ionic bond, periodic table, ion, valency, precipitate, nuclear fusion

This is not a real chemistry lesson, but what I am providing are just some basic definitions to allow understanding of what comes later.

Elements

The basis for everything on earth and in the universe is elements, which are the building blocks of everything you see around you. A few examples of elements:

C is the symbol for the element carbon

O is the symbol for the element oxygen

H is the symbol for the element hydrogen

S is the symbol for the element sulphur

Al is the symbol for the element aluminium

Na is the symbol for sodium

Cl is the symbol for chlorine

Alan McGown

Ca is the symbol for calcium

All elements have evolved since the time of the big bang, 13.8 billion years ago, where hydrogen was the first and only element in the universe. Through a process of nuclear fusion in the stars, first helium then all of the other element were formed over time.

Rocks are made of mainly silicon and oxygen. Wood, plants and animals are mostly made of carbon, oxygen, nitrogen and hydrogen. Iron ore and aluminium ore are very abundant. Water is a compound of hydrogen and oxygen, and air is mostly the elements nitrogen, oxygen and a lot if minor gases.

It may surprise you to see that the most abundant elements on the earth's crust are:

Oxygen	47%
Silicon	28%
Aluminium	8%
Iron	4%
Calcium	4%

These amounts will vary if you look at different sources of information.

Atoms

Each element has a similar atomic structure with a nucleus of protons and neutrons surrounded by orbiting electrons. The differences between different elements is that the atoms have different numbers of protons, neutrons and electrons

There are various theories as to whether the electrons should be shown orbiting the nucleus as I have shown here, or whether they just exist as a cloud.

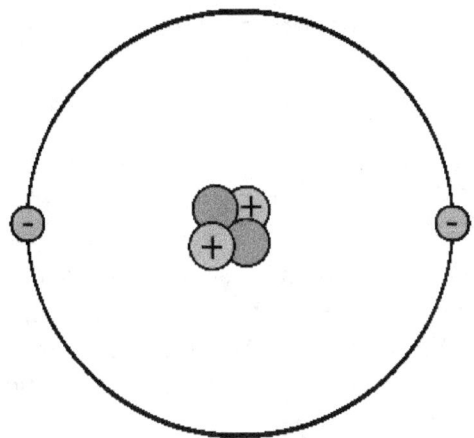

Diagram of an atom, in this case helium. The negative charge of the 2 orbiting electrons is balanced by the positive charge of the 2 protons in the nucleus. There are also 2 neutrons without charge. If an electron is removed, then the atom has a net positive charge, and if an electron is gained then the atom has a net negative charge.

A heavy atom has a large number of protons and neutrons in its nucleus, for example the heaviest natural element, uranium with atomic weight of 238 has 92 protons and 92 electrons and 146 neutrons. The lightest element, hydrogen, has an atomic weight of 1, has 1 proton and 1 electron and no neutrons. Atomic weights are found in the Periodic Table of Elements and are the basis for calculations of reactants and products in a chemical reaction.

Molecules

A combination of atoms is called a molecule. For example oxygen, O_2 (two atoms) is the form in which oxygen exists. In fact most elemental gases such as hydrogen, oxygen and nitrogen exist as molecules of two atoms.

The water molecule is H_2O. The small subscript number in the formula for water above indicates the number of atoms of that type in the molecule. A number in front of the molecule in a formula indicates the number of those molecules in the formula, for example the reaction of two molecules of hydrogen with one molecule of oxygen gives two molecules of water and heat energy.

$$2H_2 + O_2 \rightarrow 2H_2O + \text{energy}$$

This is a good example of a balanced chemical reaction. Note that there are the same number of atoms of each element on the left and the right of the arrow in the reaction. The reaction needs to be balanced if one wishes to calculate quantities of reactants or products.

Many reactions can be made to go in the reverse direction, for example in the above reaction, if energy in the form of electricity energy is put into water through two electrodes, then the water separates into oxygen and hydrogen. This is called electrolysis, and electrolysis has many commercial applications, the most significant of which is the smelting of aluminium as described in chapter 8.

Molecules consist of atoms bound together, and these atoms are held together by bonds. Compounds, molecules and reactions can be represented with or without all of the bonds being shown depending on what information is required to be shown.

Compounds

As the name suggests, compounds are composed of numerous elements, held together by bonds.

Single bonds

— Represents a single bond between atoms. Sometimes the bonds are not shown if two atoms are next to each other, such as the formula for water above, but are usually shown when describing the structure of a more complex molecule, for example the gas ethane can be shown in various ways.

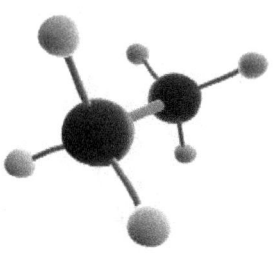

or

$$CH_3—CH_3$$

or

$$C_2H_6$$

5

Alan McGown

or

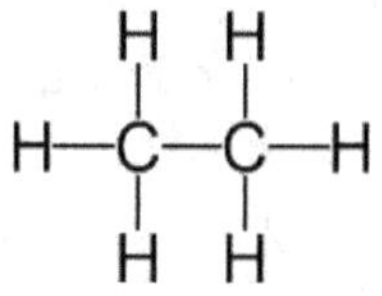

Double bonds

= Represents a double bond between atoms, for example the gas ethylene can be shown in different ways:

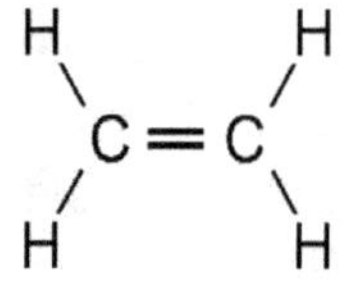

or

$$CH_2=CH_2$$

or

$$C_2H_4$$

In fact some molecular diagrams do not even show all atoms, as in benzene.

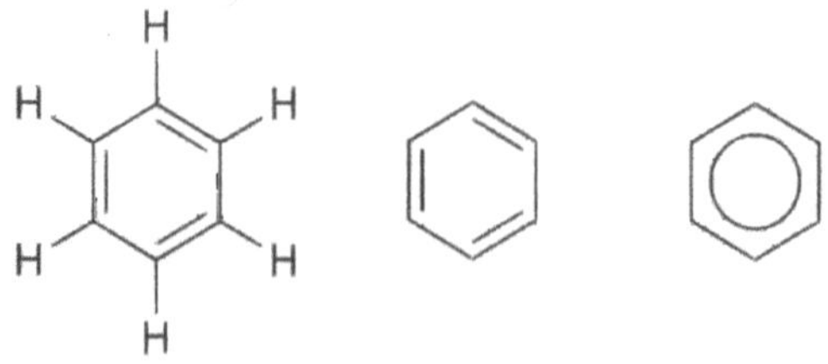

Diagram of different representations of benzene. In these examples it is assumed that there is a carbon atom at the bond intersections. The second and third diagrams assume a hydrogen atom is attached to each carbon atom. This is common when drawing benzene molecules. The circle in the third diagram is to indicate that according to one particular theory, the electrons are spread evenly around the molecule.

Hydrogen bonding

These occur in organic compounds where hydrogen atoms have an affinity with oxygen or nitrogen atoms in nearby chains of molecules. Hydrogen bonds although very weak are very widespread and a good example is in how starch is used in foods and papermaking. See chapter 13.

Organic compounds

Organic compounds contain carbon atoms in the molecule, and the atoms are held together by covalent bonds. For example ethane and benzene as shown above. For the purpose of this book I do not go into a further description of bonds.

Inorganic compounds

Inorganic compounds generally contains no carbon atoms and when solid, the atoms are held together by ionic bonds. An example is salt, $NaCl$.

Ions

When soluble inorganic compounds are dissolved in water the atoms lose or gain electrons, turn into ions and have negative and positive charges. For example salt, NaCl, when dissolved in water, turns into sodium, Na $^+$ ions and chloride, Cl $^-$ ions.

Precipitate

A precipitate occurs when dissolved chemicals are mixed with something which will cause a combination which is not soluble. The insoluble compound will turn into suspended solids which can be separated by filtration.

For example by dissolving sodium sulphate, Na_2SO_4 in water you get sulphate ions, SO_4^{2-} and sodium ions, Na^+ in solution. If barium chloride is dissolved in water it forms, barium, Ba^{2+} and chloride, Cl $^-$ ions. When these two clear, transparent solutions are put together, the barium and sulphate ions get together to make a precipitate of barium sulphate which is insoluble in water

$$Ba^{2+} + SO_4^{2-} \rightarrow BaSO_4 \downarrow$$

Barium sulphate is not soluble in water and forms a precipitate as indicated by the down arrow, which occurs as fine cloudy particles in the water. The precipitate of barium sulphate particles can be filtered out, dried and weighed. This is the basis of a classical test method for quantitative analysis of sulphate in water.

Chapter 2
Numbers, prefixes and abbreviations

We use prefixes to help state very large or very small numbers. Also some prefixes can be the same letters as abbreviations, for example m can be metres or milli, and so millimetres abbreviated is mm.

Also note that abbreviations never show plural, for example 105 millilitres of water is written as 105 ml.

A few common abbreviations are shown following. Note that some letters are in upper case and some are in lower case and are not interchangeable.

G is giga and means one thousand million, for example gigalitres written as Gl, to state the quantity of water in dams or rivers.

M is mega and means one million, for example megabytes is one million bytes used in computer terminology.

k is kilo and means one thousand, for example kilometres written as km, or kilograms written as kg, which are in everyday usage.

m is milli and means one thousandth, for example 1 millimetre is one thousandth of a metre and is written as 1 mm, very commonly used in engineering drawings.

μ is micro and means one millionth, for example 1 microgram is one millionth of a gram, used in science to describe very small weights, and micrograms is written as μg,

n is nano and means one thousand millionth, for example wavelength of light is measured in nanometres abbreviated as nm.

A few common abbreviations used in science:

Gigalitre	Gl
Megalitre	Ml
kilolitre	kl
litre	l
millilitre	ml
microlitre	µl
kilogram	kg
gram	g
milligram	mg
microgram	µg
kilometre	km
metre	m
millimetre	mm
micrometre	µm (commonly called a micron)

Some easy maths tricks

Over the years I found a few little tricks which made calculations easier. Some of these may be so obvious for some of you, so if you know it already, just pass onto the next part.

1. To divide or multiply by 10

To divide by 10 move the decimal point one place to the left and to multiply by 10 move the decimal point one place to the right. Eg 259 is actually 259.0. Divide by 10 makes it 25.9 and multiply it by 10 is 2590.

To divide or multiply by 100 move the point 2 places.

For 1000 move the point 3 places, etcetera.

2. Per means divide

For example, kilometres per hour means kilometres divided by hours. If you have travelled 250 km and it took 5 hours, then your average speed is 250 divided by 5, equals 50 km/h.

If the unit of density is grams per cubic centimetre or g/cc, how dense is a piece of stone? Say a block of stone weighs 500 grams and its size is 10cm x 10cm x 2cm, its volume is 10 x 10 x 2 or 200 cc, then grams divided by cubic centimetres, 500/200=2.5 g/cc.

In other common density units the block is 2.5 kg/1000cc or 2.5 tonnes per cubic metre.

3. Fractions, decimals and percentages

Get used to changing from one to the other in your head. For example:

One half is ½. Use your calculator, 1 divided by 2 comes out at 0.5. 50% is 50 per 100. Move the decimal point 2 places to the left and you have a decimal of 0.5. Other common ones are:

0.1 is 1/10 or 10%

0.2 is 1/5 or 20%

0.25 is 1/4 or 25%

0.75 is 3/4 or 75%

1.0 is 1 or 100%

1.5 is 1 ½ (one and a half) or 150%

3.25 is 3 ¼ or 325 %

To change a number to a percentage multiply by 100.

4. To add a percentage

Multiply the number by a number bigger than one. You could calculate the percentage then add it. For example, add 20% to 15, 20% is one fifth so find one fifth of 15=3 plus 15 =18

Much simpler to use your calculator. 20% is 0.2, so 15 x 1.2 = 18.

Another example Add 50% to $35, so 1.5 x $35= $52.5

5. To take off a percentage

Multiply by a number less than One. For example reduce the price of an article at $29 by 10%. Just multiply it by 90% or 0.9. 29 x 0.9=$26.1

6. Conversions and many calculations

Do it in easily understandable steps, rather than trying to make the calculation in one step or to remember lots of formulas.

For example I once had to find out how long it took a car to travel 1 kilometre at 100 kilometres per hour.

100km/hr = 100km in 1 hour

There are 60 minutes in 1 hour so speed is 100 km per 60 minutes.

So, time for 1 km is 60 divide by 100

Move decimal point 2 places to the left

=0.6 minutes

If you want it in seconds

0.6 x 60

=36 seconds

If you need to know how many seconds for 1 km at 80 km/h

60/80 =0.75 minutes

0.75 x 60 = 45 seconds

If you try to do it all in one go, sometimes you can't figure out when you need to divide and when you need to multiply. As long as you understand what you are doing in each step you will get to the answer you need.

7. To square a number

e.g. 4½ x 4½

Take half from one 4½ to make it 4 and add it to the other 4½ to make it 5.

4 x 5 is 20 and add a quarter. Answer is 20¼ or 20.25. Check it with your calculator.

8. To check if something is "square" or to make a right angle triangle

Once I had to set out concrete formwork for a car-port, knowing the dimensions, and I made markers for the four corners. As they were roughly a rectangle I simply checked the diagonals. When the diagonals are equal, each corner is 90 degrees or a right angle.

The second method I call the 345 rule which is better known as Pythagoras Theorem. It says that the hypotenuse squared is equal to the sum of the squares of the other two sides. The hypotenuse is the side opposite the right angle in a right angle triangle.

If you make a triangle with sides of 3, 4 and 5 units (it doesn't matter whether the units are metres, centimetres or kilometres) then 5 x 5= 25 and

3 x 3 plus 4 x 4 =25

Once you know that you don't have to worry about Pythagoras and the squares any more, just remember that the sides need to be 3, 4 and 5 units.

So I found that my corner markers were not quite square as my hypotenuse was not 5 metres. I just moved the markers until two sides are 3 and 4 metres and the hypotenuse is 5 metres.

Chapter 3
Heat and Temperature

Key words

specific heat capacity, latent heat, calorie, joule, kilojoule, calorific value, conduction, heat radiation, convection, wind chill factor, wet bulb temperature, relative humidity, dew point

There is a difference between heat and temperature. They are not the same. When heat is put into an object, its temperature rises, and when heat is taken out of an object its temperature decreases.

If you have 2 gas burners lit they are generating twice as much heat as one, but the temperature is the same, about 1000 degrees Celsius.

Heat flows from a hot body to a colder body.

Think of two saucepans of water, one full and one half full on a gas stove. When the burners are lit, heat flows from the flame, through the metal saucepan and into the water. It seems obvious that the water in the half full saucepan will come to the boil sooner than the water in the full saucepan, but why is it so?

When the water in the half full saucepan reaches boiling point its temperature will be 100 degrees Celsius while the temperature in the full saucepan may only be about 60 degrees. The same amount of heat has been put into the two saucepans but one is a higher temperature than the other. This shows the concept of what is called thermal mass. The full saucepan has more thermal mass than the half full one and it takes more heat to raise the temperature of the full saucepan.

This is because all materials have what is called specific heat capacity. This used to be described in calories where with water being the standard, it takes one calorie of heat to make the temperature of one gram of water rise by one degree Celsius.

I am using calories in this example as it helps understand the concept. These days calories are not used, the official unit of heat or energy are called joules or kilojoules.

1 calorie equals 4.2 kilojoules.

If our two saucepans contained ½ and 1 kg of water respectively, we can calculate how much heat was absorbed, assuming a starting temperature of 20 degree, no loss of heat to the atmosphere and that the specific heat content of the saucepans is not taken into account. The one kg of water has absorbed (40 degrees temperature change x 1000 grams of water) = 40,000 calories and has reached 60 degrees.

When we use that same amount of heat for the ½ kg of water its temperature has reached 100 degrees. (80 degrees temperature change x 500 grams of water = 40,000 calories).

So by this calculation the same amount of heat was consumed by each of the two saucepans of water, but because of thermal mass the larger saucepan has risen only half the temperature increase of the small saucepan.

If the two saucepans of water are left on the burners, the temperature of the 1 kg of water keeps rising and the heat being put into the half kg of water now is absorbed in making the water boil. Its temperature does not rise above 100 degrees, but the heat once the water is boiling, is now taken up by what is called the latent heat of vaporisation of the water. That heat can be given up by the steam when it condenses back into hot water at 100 degrees, as you would find out if you put say a spoon or your finger into the steam. The spoons temperature rises because it has taken the heat from the steam when the steam turns back into water as droplets on the spoon. So you can see that heat is transferable and it can be either given out or absorbed.

When your skin is wet and a breeze makes your skin cold then it is the latent heat of vaporisation being absorbed by the water turning into vapour, which makes your skin feel cold. This is also the case when the weather forecaster says that the temperature is going to be say 15 degrees and wind chill factor is going to make it feel like 8 degrees. This wind chill factor is simply the difference between a thermometer reading and the so called wet bulb temperature reading. This wet bulb temperature comes from a thermometer with a small sock of cotton over the bulb which is kept wet by dipping into a container of pure water. This difference between the two thermometers is also used to calculate the relative humidity of the air.

While we are on this subject I should explain Relative Humidity. This is the percentage of moisture the air can hold at that temperature. So at our example temperature of 15 degrees the air can hold 12.8 grams of water per cubic metre, and that air is said to be 100% relative humidity, saturated or at its dew point. Any cooler and moisture will start to deposit on a surface or in the air (fog) and at that condition the wet and dry bulb temperatures are equal.

If that air was at 70 % relative humidity then it would hold (0.7 x 12.8) 8.96 grams of water per cubic metre of air and the wet bulb temperature will be 12 degrees. Actually the Relative Humidity is calculated from the dry and wet bulb temperatures.

There is also latent heat of fusion or melting which is the amount of heat removed from a liquid once it is at the melting point to make it change from liquid to a solid. One has to put the same amount of heat into ice to make it melt. Ice in a glass of water is melting and taking heat from the water. Once the ice has all melted then the temperature of the water will stabilise. It is not because the ice is cold that it lowers the water temperature, it is because it is melting and absorbing the latent heat of fusion.

Another measure of heat is called calorific value (also measured in calories or kilojoules) which is the heat given out when a combustible material is burned. When testing a material such as coal, a small amount is weighed, then reacted with oxygen in a container called a bomb which sits in a small bucket of water. The heat produced by the burning coal heats the bomb and the water, and when its temperature rise is measured, the calorific value can be calculated in units of megajoules per kilogram or Mj/kg or calories per kg if you wish. The larger the temperature rise, the higher is the calorific value of the coal.

Coal after being mined has to have its calorific value measured so that the power station which is using that coal to boil water to make steam to turn the turbines to make electricity, will know how much coal will be consumed. The coal mine has to supply coal with certain calorific value as well as specified ash content and moisture content. Typically power station coal is required to have a calorific value of about 30 megajoules per kilogram, or Mj/kg.

Calories or kilojoules used in food and dieting are just the same heat energy as described above, and can be tested in the same calorimeter. The use of the calorific value means that when the food is digested and eventually reacted with oxygen in the body, it gives out heat to keep you warm and provide energy for your body movements.

If there is more energy produced than the body requires, then the excess is converted into fat and deposited in the body for use later on. For example when a bear eats a lot of food before hibernating, it puts on a lot of fat, and the bear can go into hibernation for months and the fat will be gradually consumed by the body to maintain body heat and provide energy for breathing.

Alan McGown

Heat transfer.

There are different ways heat is transferred from one thing to another.

1. Conduction

Everyone knows what a solid is and it depends on temperature. Temperature means that the particles of the solid are vibrating, and a body at high temperature is vibrating more vigorously than a body at a lower temperature. When a solid is heated its temperature rises and the intensity of the vibration increases. In a solid the vibrating particles are bound together and only if the vibration gets very intense the particles will not stay attached to each other, then the solid may melt and turn into a liquid.

When one hot solid (vibrating intensely) is brought into contact with a cooler solid (vibrating less) the vibration of the hotter one is transferred to the cool one. Basically we say that heat flows from a hot body to a cooler body, but it is really about the vibration being transferred. The hot body becomes cooler and the cool body becomes hotter, and in the end the temperature of the two bodies becomes equal. This is called heat transfer by conduction and can occur between solids, liquids and gases, although the different states may be separated by a heat exchanger.

For example the hot water in the cooling system of a car engine flows through the tubes of the radiator and cooler air also flows through the radiator but on the outside of the tubes in the radiator. The hot water becomes cooler and the cool air becomes hotter.

Another example is when one puts a spoon into a hot cup of coffee. The spoon becomes hot and the coffee becomes slightly cooler, but both end up at the same temperature. This is another example of thermal mass. The small mass of the spoon causes its temperature to rise a lot but the relatively large mass of the coffee cause a small drop in temperature. If you put 10 cold spoons into the cup, then the coffee becomes noticeably cooler.

2. Radiation

The heat in a body can also be lost by radiation, like when you feel the heat being given off from a fire.

Radiation is a strange thing in that the colour of the solid has an effect on the amount of heat being radiated. A matt black body radiates more heat than a shiny body and also absorbs more heat. You may have noticed the overhead radiators in some large warehouse type retailers and the hot pipes which are radiating the heat are matt black. If the pipes were shiny they would radiate much less heat. This is contrary to common sense as one would expect a shiny body to radiate more heat than a black body.

Reflection is different to radiation and everyone knows that a shiny body such as a mirror will reflect more light and heat than a black body.

Another strange effect of radiation is that radiation generally does not heat gases, only solids and to a lesser extent liquids. For example the air is not heated by the sun. The heat radiation from the sun passes straight through the atmosphere and heats the ground or the sea and the air is heated by the ground by conduction and convection.

3. Convection

A hot body can also lose heat by convection. When you heat some water in a saucepan you can see the water rising up from the bottom. What is causing this is that the water expands and becomes less dense as it gets hotter by being in contact with the saucepan. The hot water at the bottom is now is less dense that the cooler water above and the less dense water naturally rises to the top and is replaced by the cooler more dense water near the top.

This is similar to conduction as it occurs when a liquid or gas is in contact with a hot body. The gas naturally heats up, or its vibration is increased by being in contact with a hotter solid. In a gas the particles are vibrating so intensely that this vibration breaks any attractive forces which may have existed when that material was a solid or liquid.

Chapter 4
States of matter

Key words

solids, liquid, gas, plasma, solution, suspension, emulsion, colloid, Tyndall effect, Brownian motion, floc, flocculation, polyacrylamide resin, alum, saturation, nuclear fusion

You may think that there are 3 states of matter, solid, liquid and gas. There actually 4 basic states of matter and many others which are mixtures, sometimes with strange properties.

1. Solids

A solid is where the molecules are tightly packed together and are vibrating according to its temperature. When most solids are heated, the vibration of the molecules becomes so intense at the melting point that each molecule or particle is no longer bound to its adjacent neighbours and a liquid is formed. A solid may be a pure substance such as ice or sulphur or it may be a mixture for example steel alloy.

2. Liquids

A liquid is where the molecules are vibrating so intensely because of their temperature that they are not stuck together, can flow to fill a container and are held in place by gravity. Liquids may be pure elements such as liquid oxygen, pure molecules such as water or complex mixtures of molecules such as petrol. If you could see individual particles they would be whizzing around in the liquid. You generally cannot see this, but if a dry granulated solid such as sand is vibrated rapidly it is said to be fluidised and it behaves just like a liquid. Gravity affects a liquid, it is contained in an open vessel and it takes the shape of that vessel.

Many materials which melt easily can be changed from solid to liquid and back again, depending only on the temperature. This is used in many ways in industry to make goods by moulding. Thermoplastics melt when heated and after injection into a mould they will take the shape of the mould. When the mould is cooled and the plastic has turned into a solid, the mould can be opened and the solid plastic article is removed.

There are two types of liquid; polar and non polar. Polar liquids such as water or alcohols will mix together without forming separate phases. A polar liquid usually has hydrogen atom on one end of the molecule and an oxygen atom at the other end. Most organic liquids such as mineral or vegetable oils and petrol are called non polar. They will also mix together without forming separate phases. Add a little oil to water and you can see that they do not dissolve in each other.

3. Gases

When a liquid is heated to its boiling point so that the particles are flying about in all direction very intensely, the liquid will vaporise and becomes a gas. For example distillation used in oil refining separates low and high boiling point products. The crude oil is heated and the lower boiling point molecules vapourise and rise up through a column into a lower temperature condenser where they condense back into a liquid.

Many gases are pure substances such as oxygen, but many are mixtures for example, air. A gas cannot be contained by gravity alone and will take the shape of a sealed container such as a gas tank.

A gas can also be made from a liquid by reducing the pressure, or a gas can be turned back into a liquid by applying pressure. You may have heard of liquefied petroleum gas which is used as a fuel. This is usually a mixture of propane and butane and the amount of each determines whether it is high pressure gas which has more propane or a low pressure gas which has a larger proportion of butane.

In the gas cylinder the propane or butane exists as a liquid because the gas has been compressed until it turns into a liquid. Because of the pressure in the cylinder it can exist as a liquid at a temperature above its boiling point. In a gas cylinder used for barbecues, the tap at the top allows the pressure to be released and a gas flows through the valve through a hose into the burner where it is ignited. The gas cylinder contains liquefied gas in the bottom and when the gas at the top passes out through the valve the pressure in the cylinder is reduced and some of the liquefied gas in the cylinder turns back into a gas.

It does this by absorbing heat from its surroundings and you may notice when gas is being used from the cylinder, that your gas cylinder becomes cold at the bottom due to the liquefied gas absorbing the heat from the metal cylinder. In cold climates this may be a problem as the temperature may not be high enough to keep the liquefied gas evaporating fast enough to keep feeding gas to the burners. A gas is actually affected by gravity so that the atmosphere we live in does not float away into space.

Liquefied petroleum gas is also heavier than air and will act a little like a liquid and accumulate in depressions. This is why LPG is so dangerous in boats. Any gas which may come from a leak flows like a liquid into the bottom of the boat just waiting to be ignited.

All these states are reversible by raising or lowering the temperature or pressure. Most will be aware that the 3 states of $H2O$ (water, ice and steam) can be interchanged by varying the temperature. Ice melts or water solidifies at $0°$ Celsius and boils or condenses at $100°$ Celsius at normal atmospheric pressure. But if you reduce the pressure, most liquids will boil at temperatures far below their usual boiling point, and the reverse is also true in that a liquid can exist as a liquid above its boiling point if it is subjected to pressure. Later on I discuss boilers and in most cases they operate under high pressure and the steam is produced at higher temperature than steam at atmospheric pressure.

It is mostly the same for pure elements or chemical compounds which usually melt or boil at specific temperatures. Mixtures of different types of molecules usually do not have sharp melting or boiling points, but soften or evaporate over a range of temperatures.

4. Plasma

The fourth state of matter is plasma, which is where atoms or molecules are excited by very high temperature such as occurs in the stars or lightning or an electric spark. In a plasma, electrons are stripped from the nuclei of atoms and all float around freely as ions. On the sun and other stars the hydrogen molecules exist only as the remnants. The nucleus and electrons are separated and reactions take place where light elements starting with hydrogen are made into heavier elements by nuclear fusion, which gives out much energy as light and heat.

5. Other interesting states of matter

It's important to note that the following states of matter are mostly mixtures.

1. A **solution** of ionised atoms, for example when sodium chloride better known as salt, is dissolved in water, the molecules of salt NaCl are separated into ions of sodium Na^+ and chloride Cl^- .

    An ion is when the atom of sodium in this example has lost one electron and thus has a net charge of + and the chlorine atom has gained one electron and thus has a net charge of -.

2. Other solutions not ionised.

    a. A gas dissolved in a liquid, for example dissolved oxygen in water. The lower the temperature the more oxygen will dissolve.

That is why many species of fish prefer cold water as it contains more dissolved oxygen than warm water.

b. A liquid dissolved in a liquid, for example detergent dissolved in water to make dishwashing liquid.

c. A solid dissolved in a liquid, for example sugar dissolved in water.

Note 1. All liquid solutions are transparent. If it is not transparent it must be a suspension or emulsion.

Note 2. Saturation means that a certain amount of the solute can dissolve in the solvent at a certain temperature. For example copper sulphate $CuSO_4$ 5 H_2O commonly called bluestone, will dissolve in water at the rate of 20.8 grams per 100 ml at 20°C and that is called a saturated solution. That means that any more than 20.8 grams per 100 ml at 20°C added will not dissolve. If the temperature is raised to 100°C then 76.8 grams per 100 ml will dissolve.

The interesting thing is that when you have a saturated solution and the temperature is lowered the solution becomes supersaturated, the amount which can stay dissolved is less, and the excess will come out of solution as crystals. If the temperature is lowered gently the crystals will only appear if they have something to initiate the crystallisation. You can add a small pinch of small crystals and suddenly a mass of crystals will magically appear.

3. **Solid solution**, which is a mixture of solids, for example different plastics melted and mixed together above their melting point then cooled to below their melting point, or alloys of different metals made in the same way.

4. **Dispersion** of liquid in solid, for example, water in clay. This has the interesting property of plasticity which means that it can have its shape changed by exerting force. It can appear to be a solid but when put under pressure its shape can change. A common example is plasticine or modelling clay.

5. **Dispersion** of solid in liquid, for example, clay particles suspended in water to make mud.

6. **Colloid**. A colloid is a type of dispersion where solid particles are suspended in a liquid, but because they are so tiny they do not settle out.

7. **Dispersion** of solid in gas, for example dust

8. **Dispersion** of liquid in gas, for example, smoke, mist, fog, clouds, technically called aerosols.

9. **Dispersion** of gas in liquid, for example, foam or froth.

10. While at this stage, I need to mention **flocculation**, which means converting the very small dispersed particles of a suspension into large ones so that they can be filtered or settled out of the suspension. Water treatment at domestic water purification plants uses synthetic water-soluble so called polyelectrolyte resins or simpler chemical such as aluminium sulphate (alum). Alum reacts

with alkaline groups in water to form aluminium hydroxide when in contact with solid particles, the aluminium hydroxide clumps together to form a floc and this then settles carrying down the suspended contamination particles. See chapter 13 where alum use in papermaking is discussed.

11. **A gas** mixture such as air, which is mostly nitrogen, oxygen, carbon dioxide and very small amounts of many others such as hydrogen, helium, argon etc.

12. **A gel** is a type of solution where the molecules are very long and interact with each other. A good example is starch. See it in chapter 13. When the starch has been cooked, the very long molecules can be subject to hydrogen bonding which means that when the molecules are in the form of chains there is a weak bond formed between hydrogen atoms and oxygen atoms. Only when the concentration is high enough for the molecule chains to get close to each other this can happen, the chains are attracted to each other and seem to stick together and the solution can be almost solid. This is what causes lumps in custard or gravy. The lumps are areas where the concentration of the starch is highest.

13. **Emulsion.** Dispersion of a liquid in liquid called an emulsion. For example, milk, which is a complex emulsion having liquid fat and solid protein dispersed as small globules in water. There are two types of emulsions and both have polar phase and a non polar phase.

a. **Oil in water**, an example is milk and cream. Non polar dispersed phase (milk fat) exists as globules and polar continuous phase (water). When some salt is added and the cream is churned with high shear it reverts to butter which is a water in oil emulsion.

b. **Water in oil**, an example is margarine or butter where the water exists as minute globules dispersed in the oil. Polar dispersed phase (water) and non polar continuous phase (oil).

There are many emulsions in industry for example oil based cutting fluid used to lubricate during metal drilling and machining. Water based paints are complex emulsions of resins, solvents and pigments.

Many paints have another interesting property called thixotropic. That means they are relatively thin or low viscosity when mixed in the can and when being painted onto a surface, but become thick or high viscosity once applied to the surface. This allows thick layers to be applied without the problem of slump when the surface is vertical, which was always a problem with older paint formulations.

Now we get to the different industries in which I was employed.

Alan McGown

Chapter 5
Aluminium foil manufacture

Key words

aluminium foil, anneal, grain shape, stearic acid

My first job was to test aluminium foil, which was produced by cold rolling aluminium sheet which was supplied in large rolls from the company's other factory. The rolling mills were large machines with two rollers about 200 mm diameter and about 1500 mm wide, The rolls were machined incredibly accurately as the thickness of the foil could be as low as 9 microns which was half the thickness of the gap between the rolls. I'll explain why shortly. The surface was also very smooth as the smoothness produces the bright smooth finish on the surface of the foil.

Aluminium foil is made with one side bright and the other side matt by a simple process. Two layers of aluminium sheet are fed through the rolls and the foil produced is separated after rolling. The outside which was against the very smooth rolls comes out bright and shiny and the inside against the other piece of foil comes out matt when separated.

Rolling oil was applied to the rolls to lubricate and cool the rolls and foil, and part of the laboratory assistant's job was to also test the oil. Aluminium sheet about 1mm thick on a roll is made by cold rolling aluminium in several stages. The sheet is on a roll about 1 to 2 metres wide and several hundred metres in length.

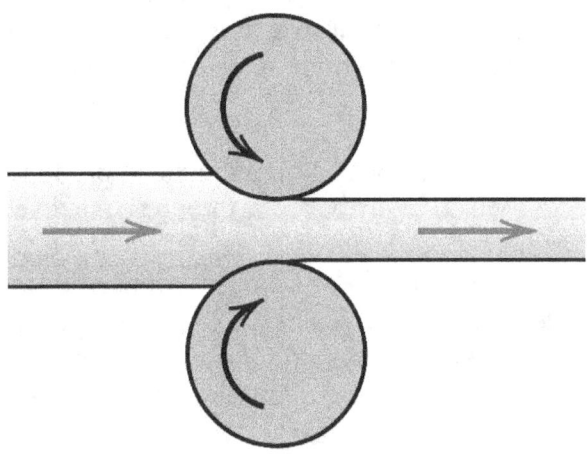

Diagram of rolling aluminium sheet into thinner sheet. The same process is used to roll aluminium sheet into foil.

Rolling produces a foil which is hard and the grains inside the metal are elongated. Most foil needs to be soft to perform its functions and that is achieved by annealing the foil. A large roll of foil is placed in a large oven and heated up to about 290 to 400 degrees Celsius depending on the different alloys, then allowed to cool slowly.

When the aluminium has cooled, the metal grains have changed from elongated to shapes as shown below and the foil is now soft.

Alan McGown

My job was to take samples of foil to check that they were properly annealed. This was done by etching the surface with aqua regia, which is a mixture of nitric and hydrochloric acid. Aqua regia (royal water) is just about the strongest acid available. It will dissolved metals such as gold which cannot be dissolved by any other acid.

The aqua regia produces on the surface of the foil, a grey area which when looked at through a microscope, shows the individual grains of aluminium. If they were elongated then it was not properly annealed, and if annealed properly the grains were roughly circular but uneven. Different alloys produced grains of different size and using the microscope we measured the grain size.

Piece of aluminium which has been etched to allow the grains to be seen. This piece is soft as you can see that the grains are not elongated.

The foil was produced in different thicknesses for various applications. The thinnest grade usually 9 microns, less than half the thickness of a human hair, was used for laminating with very light paper for lining cigarette boxes.

Another grade of about 30 microns thick was used for making moulds for apple and meat pies. This grade also had a food grade lubricant applied, and this was food grade stearic acid dissolved in a petroleum solvent. The lubricant was applied to the surface of the foil so that the machine producing the moulds could punch them by using a male and female die and could run at a high rate of output without jamming.

Stearic acid is a fatty acid derived from animal fat by reaction with sodium hydroxide and has the formula:

$$CH_3- (CH2)_{16} -COOH$$

or

$$C_{18} H_{36} O_2$$

See chapter 9, triglycerides.

Another grade of foil was called milk strip. In the 1950-1960's milk was sold in glass bottles with a foil cap. The thickness of the foil was about 30 microns and often was printed with the milk producers name or other decoration. The foil was slit into widths of about 45 mm wide and the rolls were about 200 mm in diameter.

Foil printing

The printing was another process, where coloured clear or opaque lacquer was applied to the full width foil rolls before slitting into different widths required by the customers. The lacquer was applied by a large printing press with either metal or rubber rolls. The pattern required was etched into the printing roller which turned in a bath of lacquer and the lacquer was picked up by the roller. Excess lacquer was squeegeed off the smooth surface of the roller by a "doctor blade", leaving lacquer in the etched surface. Then when the roller was pressed against the foil, the lacquer in the etched surface was left on the foil, or as they say in the trade was "offset" to the foil. The foil was then heated with hot air to evaporate the solvent from the lacquer and leave a clear dry layer of lacquer.

Clear lacquer was used and it is also coloured by adding dyes or opaque pigments to make whatever fancy patterns the customers want.

The lacquer was mostly cellulose nitrate, also called gun cotton, as it was the basis of explosive propellants in large artillery shells. The cellulose nitrate was dissolved in a mixed solvent of methyl ethyl ketone, acetone, ethyl alcohol and ethyl acetate mixed in suitable ratios to give a desired evaporation rate when passed through the hot air drying section of the printing press.

Other grade of foils were used for sealing plastic food tubs, for example yoghurt and other foods. These grades usually had fancy pattern printed on the upper side and the clear lacquer on the underside was based on vinyl lacquer so that it could be heat sealed onto the plastic tub.

Foil for tablet packs had the same vinyl heat seal lacquer printed onto one side so that two layers of foil could be used to sandwich the tablets.

Chapter 6
Oil refining and petrochemicals

Key words

crude oil, paraffin, olefin, isomer, distillation column, pyrolysis, gas, coal tar, pitch, polymerisation, styrene, aromatic, water tube boiler, fire tube boiler, catalytic cracker, hydroforming, alkylation, gas chromatograph

Crude oil consists of a wide range of organic compounds, and the composition varies according to where the oil is found. Middle East crudes are of high viscosity (thick) and contain high molecular weight compounds. Australian Bass Strait crudes are composed of lower molecular weight compounds and are low viscosity. Straight from the ground crude oil and gas contains:

Chemical name	Formula	No. of carbon atoms
Hydrogen gas	H_2	0
Hydrogen sulphide gas	H_2S	0
Methane gas	CH_4	1
Ethane gas	$CH_3\text{-}CH_3$	2
Ethylene gas	$CH_2=CH_2$	2
Acetylene gas	$CH\equiv CH$	2
Propane liquefied gas	$CH_3\text{-}CH_2\text{-}CH_3$	3

Chemical name	Formula	No. of carbon atoms
Propylene gas	$CH_3-CH_2=CH_2$	3
Butane liquefied gas	$CH_3-2CH_2-CH_3$	4
Pentane liquid	$CH_3-3CH_2-CH_3$	5
Hexane liquid	$CH_3-4CH_2-CH_3$	6
Heptane liquid	$CH_3-5CH_2-CH_3$	7
Octane liquid	$CH_3-6CH_2-CH_3$	8
Nonane liquid	$CH_3-7CH_2-CH_3$	9
Decane liquid	$CH_3-8CH_2-CH_3$	10

Etc etc up to about C26 (26 carbon atoms) or higher depending on the source of the crude oil.

The gases mentioned can come from the ground either in the gaseous form without any oil, or the gases may be dissolved in crude oil. These are called alkanes where there are no double bonds, and are also called saturated, meaning that no more hydrogen atoms can be attached. Alkenes have one or more double bonds for example ethylene and propylene and these are called unsaturated as more hydrogen atoms could be added at the double bond.

It is the double bonds which allow polymerisation, which is the joining together of large numbers of the monomer into polymers or plastics, like polyethylene.

As mentioned before, all molecules are actually 3 dimensional. Compare the formula of n-octane with its 3D representation.

CH$_3$-6CH$_2$-CH$_3$

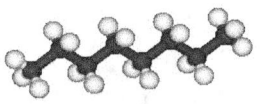

3D diagram of normal or n-octane showing the black balls as carbon backbone and the attached white hydrogen atoms.

Waxes, commonly called paraffin wax (C20 to C40) are long straight chain organic chemicals which come from crude oil, in fact they are a nuisance to refiners, as a crude oil containing lots of wax will set almost solid when it is cool.

Paraffin wax is not related to other waxes such as beeswax or waxes from plants such as carnauba wax. De-waxing of crude oil is a process where the wax is allowed to solidify while the whole mass is mixed vigorously to form a slurry and then this is passed through a large filter. Sometimes these waxes are left in the oil and the whole liquid is kept warm if the material is used as a feed for "cat cracking", hydrocracking, or hydroforming.

Isomers are different structured molecules with the same number of carbon and hydrogen atoms. With each increasing additional carbon atom in the molecule, the number of isomers increases. The simplest example is butane, C$_4$H$_{10}$ which takes 2 forms:

Normal or n-butane:

CH$_3$-CH$_2$-CH$_2$-CH$_3$

Iso butane:

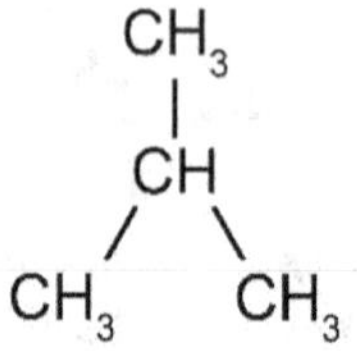

$$CH_3$$
$$|$$
$$CH$$
$$CH_3 \quad CH_3$$

Unsaturation or double bonds may also appear in different places in the molecules, so that again in the simplest form butylene or more correctly named butene, appears in 3 forms of C_4H_8:

Normal butene or 1-butene where the double bond appears between carbon atoms 1 and 2:

$$CH_2=CH-CH_2-CH_3$$

2-butene where the double bond appears between carbon atoms 2 and 3.

$$CH_3-CH=CH-CH_3$$

and iso butene which is branched

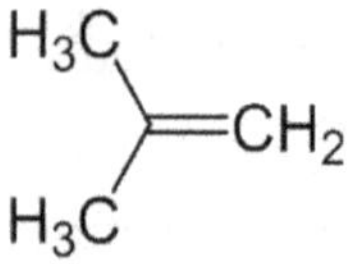

$$H_3C$$
$$=CH_2$$
$$H_3C$$

And in addition, there is butadiene C_4H_7 with 2 double bonds:

$$CH_2=CH-CH=CH_3$$

So in addition to the number of carbon atoms allowing different isomers to exist, we have the complication of double bonds which may appear in numerous places. Those based on butane are only the simplest cases.

Imagine how many different isomers may exist when the number of carbon atoms increase to the twenties.

So you can see that mineral oils are composed of millions of different but similar molecules, called hydrocarbons, because they are composed mostly of carbon and hydrogen, and each has its own unique boiling point. These can easily be separated into different fractions by distillation. Each distilled fraction will still contain many thousands of different but similar molecule types, but each separate fraction has its own individual uses determined mostly by the range of boiling points of the constituent molecules. For example:

1. Hydrogen, methane and ethane are used for town gas.

2. Ethylene and propylene are important raw materials for plastics.

3. Propane and butane are the major constituents of LPG (liquefied petroleum gas).

4. Pentane (C5) and up to heptane (C7) are the main constituents of gasoline or petrol.

5. Octane (C8) and up to dodecane (C12) are the main constituents of jet fuel, kerosene and diesel fuel.

6. C12 to C20 hydrocarbons are used as heavy diesel fuel in large ship engines (called bunker oil) and above C20 are used as lubricating oils.

Alan McGown

Distillation

This is the method used in refineries of separating the organic molecules on the basis of boiling point.

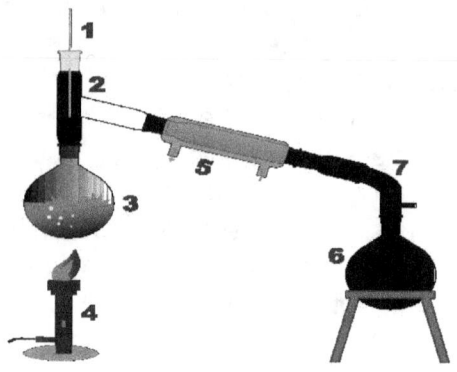

Diagram of laboratory distillation apparatus. 1. thermometer, 2. distillation adapter, 3. Flask holding the liquid being distilled, 4. bunsen burner, 5. water cooled condenser, 6. receiver flask, 7.vacuum adapter

A simple laboratory distillation of a petroleum liquid using apparatus as above, causes the lowest boiling point compounds to evaporate first, pass the thermometer, and then pass to the water cooled condenser, where the gaseous fraction condenses back into a liquid. As it passes the thermometer the petroleum is starting to condense and so the thermometer shows the boiling point of that fraction.

As each fraction of petroleum is evaporated the temperature in the distillation flask gradually rises. The sample can be split into numerous fractions, each with a higher boiling point than those before. By changing the receiver flask and noting the temperature showing on the thermometer and the volume received, the various fractions can be graphed using volume versus boiling point.

Depending on the sample being distilled, the temperature in the distillation flask will start to reach a point where the contents are starting to carbonise rather than distil. At this point the distillation is usually terminated.

The distillation can also be carried out under vacuum, which lowers the boiling point of the fractions and the high boiling point fractions can be distilled at lower temperatures, thus lessening the carbonising influence of the high temperatures.

Industrial crude oil distillation is usually the first process the crude oil undergoes at the oil refinery. The crude oil is heated and the lowest boiling point products are vaporised first and then pass upwards through a column. In simple terms this column has a range of temperatures from top to bottom and at the highest level with the lowest temperatures the lowest boiling point products are condensed from the vapour into a liquid. There are many levels each with a different temperature and each level condenses a different liquid product called a fraction.

Because there are thousands of different molecules present in the crude oil, each with a slightly different boiling point, each fraction is still composed of many similar but slightly different molecules. For example if you look at the gasoline fraction shown above with molecules from C5 to C8, because there are many different isomers there are still hundreds of different molecules. In fact in that fraction there are also small amounts of molecules on either side, C3, C4, C9, C10 etc. See the section on later pages showing gas chromatography of oil fractions.

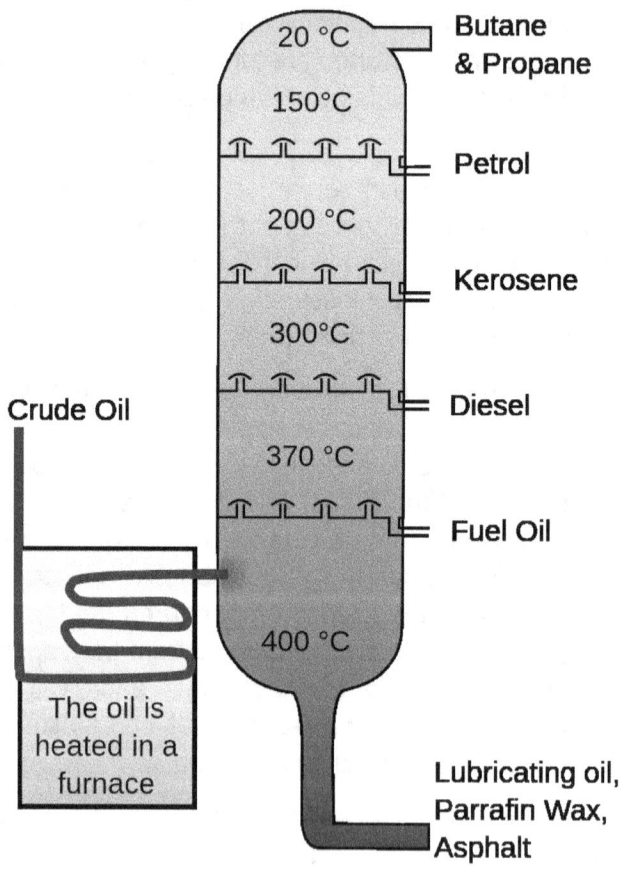

Diagram of the principle of distillation in an oil refinery. At each level a different product is being condensed.

Most of the towers in this oil refinery are distillation columns. This photograph and the diagram above are good examples that a process may be simple in basis, but is usually much more complex in practice.

The residue which has not been vapourised is left in the bottom of the distillation column and is called residual oil or bottoms. These very heavy or high boiling point residual oils are used for various purposes including being "cracked" into lower molecular point compounds by adding hydrogen at high temperature and pressure with a catalyst. This process is sometimes call "cat cracking", hydrocracking, or hydroforming.

In the 1940s to 1980s there was a lot of residual oil, more than could be used or sold as products, and it could also be used for firing directly in furnaces in the oil refinery to raise steam to be used in the many processes.

Plastics such as polyethylene are quite close in chemical structure to paraffin waxes from petroleum. Because of the similarity, the two can be mixed hot and the polyethylene actually dissolves in the wax. Such waxes were used to make corrugated cardboard cartons waterproof as the mixture was more durable than just paraffin wax alone.

Petrochemical manufacture

Pyrolysis or thermal decomposition means heating something to a high temperature in the absence of oxygen. The molecules break down into smaller molecules and when cooled they reform into different molecules. Following is a description of pyrolysis in an industrial process.

Most organic molecules will pyrolyse. For example, wood which consists of cellulose and lignins, is pyrolysed when heated by a flame. In fact when wood is burning, it is the gaseous pyrolysed molecules which are burning. Have a close look at a piece of wood burning and you will see a hole in the side of the wood and a jet of gas coming out which is what is burning. The heat of the flame is pyrolysing the wood into flammable gases which do not burn until they pass out of the wood and come into contact with oxygen in the air. When you blow out the flame you will see the pyrolysed gas coming out as white smoke and it can be relit with a match just like a gas.

An old process to make town gas, organic liquids and tar, was to pyrolyse residual oils from the oil refining process. This residual oil was a by-product of the initial distillation of crude oil and was the high boiling point oil left over after the lower boiling point fractions had been distilled out and in those days was an unwanted by-product.

I was a chemist in the laboratory at a plant which converted this otherwise unusable residual oil from refineries, into useful products by pyrolysis, also called thermal cracking. The residual oil, generally above C20 (organic molecules with 20 carbon atoms) was sprayed onto red hot bricks in a retort in the absence of air. The bricks had been heated by burning some of the oil in the retort until the bricks reached red heat, about 1000 C. Then steam was injected into the retort to drive out any air, following which the oil was sprayed onto the red hot bricks.

As there was no oxygen present in the retort, the oil did not ignite but the molecules were broken up in the process called pyrolysis. The cloud of hydrocarbon vapour was then allowed to cool, which caused the molecule fragments to reform, then the whole lot was quenched in water.

The resultant mixture of molecules now consisted of every possible combination of hydrogen and carbon between hydrogen and carbon, for example:

1. Low molecular weight hydrocarbons from hydrogen, methane, ethane and propane which was used for town gas after the ethylene was extracted to make styrene and polyethylene.

2. Slightly higher molecular weight hydrocarbons from propane to pentane (C3 to C5) which was compressed to make Liquefied Petroleum Gas (LPG)

3. A small amount of hydrocarbons up to C8.

4. Significant amount of so called "aromatic" oils based on benzene, called BTX which stands for benzene, toluene and xylene. Aromatic in this chemical sense always means based on benzene molecules. Benzene is C_6H_6 in a hexagonal shaped molecule. Toluene has one CH_3 group attached to the benzene ring and Xylene has two CH_3 groups. As there are 6 carbon atoms in the benzene ring, these two CH_3 molecules can be in 3 different places and this is another example of 3 isomers with the same chemical formula. This means that there are 3 different isomers of xylene. These were separated by distillation and sold for use as industrial solvents and the benzene along with the ethylene mentioned above was used to make styrene, an important chemical used for making a wide range of plastics and rubbers.

Diagram of benzene, toluene and 3 xylene isomers. This is a good example of isomerism. The 3 isomers of xylene have exactly the same empirical formula, ($C_8 H_{10}$), but different location *of the methyl groups (CH₃) on the benzene molecule. Note that in this case the six carbon atoms and the hydrogen atoms of the benzene rings are not shown but inferred.*

5. Then came a range of slightly higher molecular weight aromatic oils (based on the benzene molecule) containing naphthalene, anthracene and many other aromatic molecules based on the benzene molecule. These had little commercial use and were generally burnt in the boilers to raise steam used in the refinery for many distillation processes. Some were used to add to tar for road making.

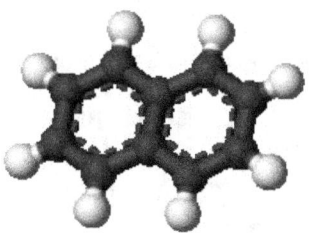

3D Diagram of naphthalene molecule

6. Next was a range of tars which are simply even higher molecular weight aromatic hydrocarbons. These were used for making roads and in particular tarmac at airports. Tar rather than bitumen was preferred for airport tarmacs as it was more resistant than bitumen to spilled straight chain hydrocarbon based fuels used at airports. Bitumen is dissolved by the fuel and the tarmac would quickly break down if made of bitumen.

Tar is also turned into pitch by distilling out the oils mentioned in points 4 & 5 above. So pitch is simply the higher molecular weight fraction of the tar and is in the form of a hard, easily melted black substance. A major use for pitch is as an electrode binder in carbon electrodes for the aluminium industry, which I will describe shortly.

7. Carbon is also left in the last fraction.

All of these above products are almost identical to the range of products which result from coking of coal for the steel industry. The process is basically the same as described above, in that coal is heated in the absence of air (pyrolysed), which causes all of the volatile products to be distilled out of the coal, which then leaves coke which is basically carbon.

A similar range of chemicals is produced and tar is the last after distillation of the other useful products out of the resulting liquid. Coal tar pitch is made in the same way as described above.

Coal tar is not used much in road making these days as it has been largely replaced by bitumen which has different properties, but coal tar pitch is still used to make electrode binders for the aluminium industry.

It is interesting that tobacco smoke or wood smoke has a very similar composition to the coal tar gas and contains lot of poisonous aromatic (benzene derivatives) compounds due to the pyrolysis. In fact smoke from bushfires or burning wood is of similar composition and is in the form of really tiny globules called aerosols. Smoke is the result of pyrolysis with insufficient oxygen to completely react with carbon based products. If there was sufficient oxygen to react, the products of combustion would be carbon dioxide and water and no smoke.

You can see this in a wood burning heater. If the air inlet is restricted, the fire produces white smoke, and tar as the tiny globules will condense on the inside of the flue pipe and eventually restrict the flow. If the air inlet is not restricted then combustion can be more complete, and will produce the maximum amount of heat, to produce carbon dioxide and water, and almost no smoke is produced.

A couple of interesting matters arose from my employment in that industry. Diesel engines will run on almost any oil, as we used to feed 3 very large diesel ships engines with residual oil, such as was used for the feedstock to the pyrolysis process. These large diesel engines with 12 cylinders and of dimensions approximately 8 metres in length 1.5 metres wide and 3 metres high were used to power the gas compressors feeding town gas to the customer, AGL.

Boilers for raising steam

Oil refineries need a source of steam to heat chemical processes and to power steam turbine pumps. Large steam boilers of which we had 4, will run well on almost any flammable liquid, solid or gaseous fuel. Again we used the residual oil to fire these boilers, but sometimes we had an excess of tar and it was also used to fire the boilers.

Alan McGown

At the refinery we had three large ships boilers which were of the water tube type. With a water tube boiler, the boiler is just a large box with insulated metal sides. See the diagram of a Yarrow boiler. At the bottom of the fire box are two tanks, one on each side, into which the cold water is fed and with steel tubes protruding from the top. These steel tubes with small spaces between pass upwards and terminate in the steam drum which is a single tank and is kept about half full of the superheated water and steam. The steam is drawn off from the steam drum.

The liquid fuel we used was injected into a space surrounded by the tubes containing the water to be boiled, and a large flame was maintained. Steam to atomise the fuel was injected through the injector, which was basically a large pipe with a series of holes on the end inside the boiler. A spray of finely divided fuel was spraying out of the end of the injector and was ignited into a large flame. The heat and gases from the flame passed upwards around the small spaces surrounding the water filled tubes and the gases exited the boiler through the top of the boiler and passed upwards through the stack (smokestack not shown). The tubes were heated by radiation from the flame and also by conduction from the hot gases passing from the flame.

Around the fuel injectors is also fed a stream of air to provide more than enough air for complete combustion, and most boilers operate on this so called "excess air" in order to get the most heat possible out of the fuel. When you see smoke coming out of the stack it generally means that the atomisation of the liquid fuel is faulty and sometimes unburnt fuel is falling onto the bottom of the boiler where it is burning with insufficient air, which produces a smoky flame.

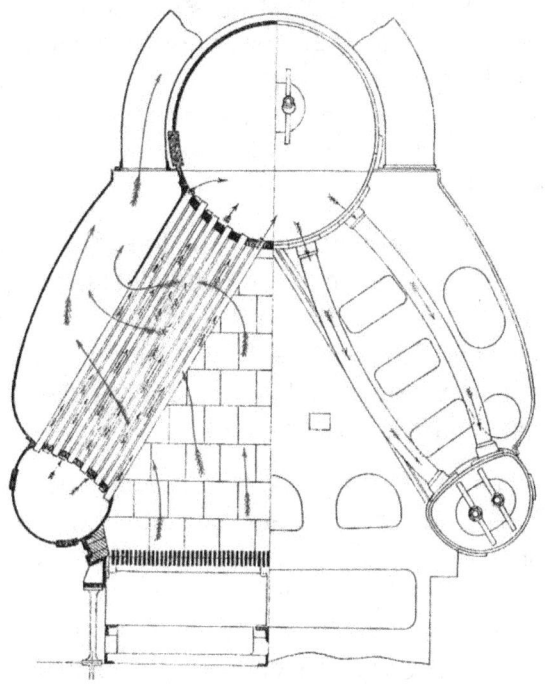

Diagram of a large Yarrow boiler. This water tube boiler was recovered from a World War 2 destroyer. There many different arrangements of industrial boilers, but all have the steam drum at the top. The hot flue gases from the burner passes between the tubes and out through the top before existing from a smokestack. The burners and smoke stack are not shown. This view is from the front where the burners are placed. Width is about 6 metres, height is about 12 metres and the depth can be around 6 metres.

With water tube boilers, cold water is fed into the tubes at the bottom of the boiler and the temperature of the water within the tubes increases as the water passes upward. The tubes all enter the steam drum which is half filled with water well above boiling point, and the steam at high pressure is taken from above the water level in the steam drum. The steam is then piped away through insulated pipes to where it is used to drive steam turbines or other types of pumps.

The pressure is proportional to the temperature of the steam. If there is too much demand for steam from the turbines connected to the steam pipes, then the pressure and temperature of the steam decreases. This can be compensated by increasing the amount of fuel being injected into the boiler.

Most modern boilers are water tube type where the water being turned into steam is inside the tubes and the flame is on the outside of the tubes. Old steam locomotives were generally of the fire tube type where the flame and exhaust gases passed through the tubes and the water was in a tank on the outside of the tubes.

Boiler water testing

Part of my job as a chemist was to test the boiler water so that the boiler could continue to operate. All water contains dissolved minerals such as salt, calcium and magnesium. As the water is continuously evaporated into steam, the dissolved minerals are left in the water remaining in the boiler and the concentration gets higher and higher. These minerals are deposited onto the insides of the tubes in the boilers as scale and can cause problems.

As the dissolved salt and other minerals from the water deposits scale on the inside of the tubes, it forms an insulating layer. As the deposit does not conduct the heat as rapidly to the water inside the tubes as do clean steel tubes. The outside of the tubes in contact with the flame gets hotter and can even reach red heat. The water inside the tubes is supposed to be keeping the steel tubes relatively cooler. What was a frequent occurrence in the past was that due to the high temperature of the tubes caused by the insulating deposits on the inside of the tubes, the tubes can eventually rupture and an explosion results with the release of all of the superheated water and steam.

Boiler water treatment is a science where the concentration of the various types of dissolved and suspended minerals in the water is tested and modified by adding chemicals, to produce a suspended precipitate which can be removed by a process called a blow-down. This is where the water in the tubes along with suspended minerals is released, by opening valves connected to the bottom tanks, resulting in a volume of hot water above boiling point being released along with the suspended minerals. pH is also tested in the boiler water to keep it slightly alkaline, above pH 7, to reduce corrosion of the steel boiler tubes.

Chemical testing in the oil industry

The processes in the oil refinery are these days computer controlled with many sensors (temperature, pressure, flow meters) and pumps valves monitored and adjusted by the operators to produce the required products in the quantities required. There are also many laboratory tests required to confirm the physical and chemical properties required to provide quality control of the products and intermediate products. I will list a few:

Gas chromatography

Gas chromatography is used to determine the chemical constituents in complex mixtures. This is almost like a distillation process. The product being tested is injected into a long thin stainless steel, glass or Teflon tube called a column, typically between 1 and 5 mm in diameter and various lengths possibly 4 or more metres long. The column is coiled in an oven which can be operated at various temperatures and also typically the temperature can be programmed to be raised at a rate depending on the type of sample being tested. At the beginning of this column is a heated section where the product being injected is vapourised. The vapour is passed through the column which is filled with a powdered solid coated with a specific chemical according to the type of product being tested.

The higher molecular weight products spend more time than the lower molecular weight products dissolved in the chemical coating the powder and are held up in the column longer than the lower molecular weight products. What happens is that by the time the first low molecular weight products are exiting the column they have been separated from the higher molecular weight products and exit the column one by one.

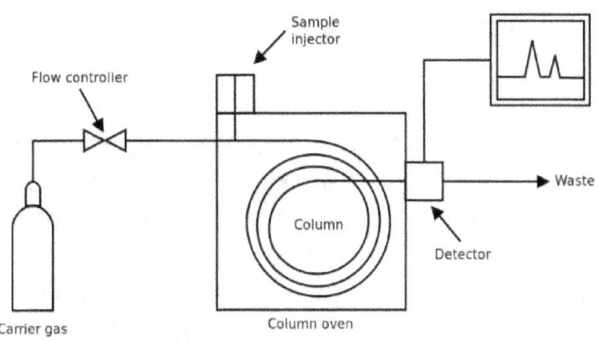

Diagram of a gas chromatograph showing only the main parts.

A detector of which there are many types, detects the products as they exit the column one by one and creates a signal which is connected to a printer which draws a graph of time versus the quantity of the products exiting the column, and looks like a series of triangular peaks where the size of the peak is proportional to the amount of each chemical from the original mixture.

Even a simple product, say propane, will still contain many related alkanes and alkenes, and its analysis may be something like:

Methane   5%

Ethane   10%

Ethylene   2.5%

Propane   65%

Propylene 2.5%

N Butane 10%

Iso butane 5%

Pentanes traces

So when a mixture like this is tested in a gas chromatograph, those individual components exit the column in order of boiling point, through a detector and each compound produces one triangle shaped peak, (really a bell curve or normal distribution curve) to be drawn by the printer. The area of each is proportional to the quantity exiting the column and the percentage of each in the original sample can then be calculated. We used to do it by physically measuring the dimensions of each peak with a pencil and ruler or even by cutting out the paper peaks and weighing each. Later printers came equipped with integrators and nowadays a computer receives the signals from the detector and calculates the percentages of the constituents.

These days a mass spectrometer is used to identify and quantify individual chemical compounds as they exit the column.

Each product to be sold will have a specification stating the maximum permissible percentages of the various components, depending on the intended end use and the capability of the separation processes used in the refinery. This is a rather simple example due to the small number of possible isomers of the various hydrocarbons.

All of the other hydrocarbon products will have similar but much more complicated mixtures, due to the large number of possible isomers with similar boiling points.

Distillation is the most common method of separating hydrocarbons in oil refineries and the separation is based largely on differing boiling points of the hydrocarbons.

It is rather interesting that the pyrolysis product from the retort mentioned above when injected into a gas chromatograph, produces a trace of many thousands of different compounds most of which are aromatic (based on benzene rings) and if cigarette smoke is also tested, it produced an almost identical trace. This is a simple example of pyrolysis of the tobacco producing harmful aromatic chemicals which do a lot of damage when ingested into the lungs of a smoker.

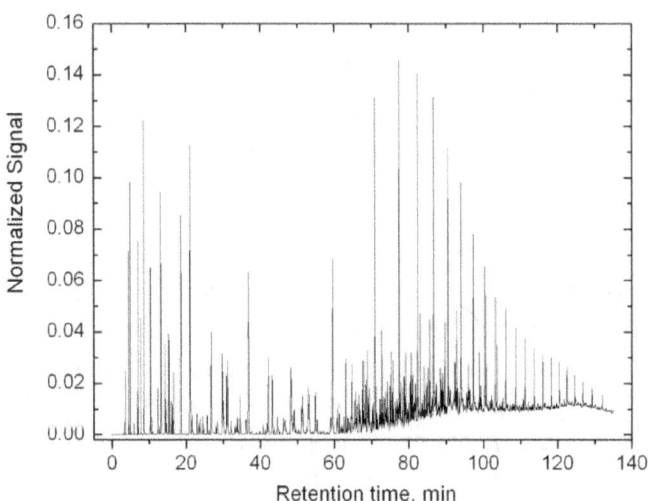

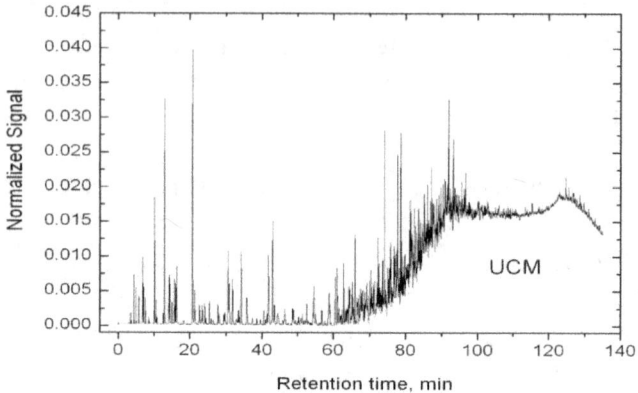

Diagram of chromatograms for 2 different oil samples (this page and prior). Note the large number of chemical compounds in the two oils. Retention time is the time between sample injection and appearance of each compound. Retention time is typical for each compound and the type of column. Both samples show a large peak at about 22 minutes and it is the same compound, showing how gas chromatography can be used for identification as well as quantitative analysis.

Calorific value

Calorific value is required to be measured for town gas products. A quantity of the gas is burned in a special burner where all of the heat passes to a small stream of water flowing through the calorimeter. The temperature increase of the water is proportional to the quantity of the gases used and its calorific value. In the 1960s calorific value was expressed in calories where one calorie is the amount of heat required to heat one gram of water by one degree Celsius and was shown as calories per cubic metre. As town gas contains hydrogen, methane, ethane, ethylene and propane, its calorific value is highest when the hydrogen content is high.

Town gas made from coal gas can also contain significant amounts of carbon monoxide and hydrogen sulphide and that is why town gas of those days was so poisonous. Both are quite flammable. Town gas these days is mostly natural gas (from coal seams) and can contain many of the same constituents.

Water gas which was quite common in gas plants up to about the 1960's is made from carbon and water. Coke (carbon) is made from coal by heating to drive off the volatile compounds without allowing it to catch fire. The coke is burned in air to red heat, then steam is injected and the reaction produces a gas which is a mixture of hydrogen and carbon monoxide.

The reaction is:

$$C + H_2O \rightarrow H_2 + CO$$

Carbon plus steam $\rightarrow$ hydrogen plus carbon monoxide

Both hydrogen and carbon monoxide burn readily in air, the reactions are:

$$2H_2 + O_2 \rightarrow 2H_2O$$

$$2CO + O_2 \rightarrow 2CO_2$$

Both gases are quite flammable and carbon monoxide is particularly poisonous, and that is the reason why town gas was known to be poisonous. It is a colourless, odourless, tasteless gas which can be breathed without being aware of it. It is poisonous as it forms a stronger bond with haemoglobin in the blood than does oxygen. The symptoms are dizziness, unconsciousness and death, depending on how much was breathed. Town gas these days or liquefied petroleum gas LPG is not particularly poisonous but as it is denser than air, it is as asphyxiant, that is it displaces air and one can be asphyxiated due to lack of oxygen.

Odourants are put into town gas and LPG so that the gas can be detected by smell. These odourants are generally sulphur containing molecules such as mercaptans which can be detected by smell at extremely low levels and will act as a warning of gas leaks. The smell is like garlic.

Density

Density of many products is measured by various methods. As density is simply mass per unit volume, one only has to measure mass and volume. For liquids a hydrometer is used which is a calibrated glass float which floats higher in more dense liquids. A calibrated container can be used which contains a specific volume and the mass of a liquid to fill it allows the density to be calculated.

For solids such as pitch a common method is to weigh a lump of the product in air and in water. The loss in weight when weighed in water equals the volume of the piece and the density calculation is the mass divided by the volume.

Viscosity

Viscosity of tar, bitumen and oils is measured by various methods. In the simplest method, the melted tar or bitumen at a specific temperature is poured into a small cup with an orifice in the bottom blocked by a plug. The time is taken for the quantity of product to flow out of the cup.

Lubricating oil is graded according to viscosity and it is measured in the same way by passing a known quantity through an orifice in a glass u-tube mounted in a temperature controlled water bath with glass sides.

Viscosity or thickness is a very important property of lubricating oil and it is further complicated by viscosity reducing with increased temperature. An oil of lower viscosity lubricates less well than an oil of higher viscosity and in the engine of a car this can lead to damage if the engine is running at a high temperature. But in winter time the oil may be of too high a viscosity to pass rapidly through the many small passages in an engine before it warmed up and "cold start wear" could be a problem.

In the 1950's, engine oil used to be classified as Society of Automotive Engineers (SAE) ratings and typically grades were 20, 30, 40, 50 SAE. In the 1960's multigrade oils were developed to overcome the problem of cold start wear so that you could use for example 20w30. This means that this oil is a 30 SAE rating but its viscosity in winter when cold is more like 20 SAE oil.

These days a typical oil used in car engines may be a 10w50 and this is a description of the viscosity index of the oil. Viscosity index is a measure of how much the viscosity of an oil changes with temperature.

This was achieved by having viscosity index modifiers added to the oil to reduce the loss in viscosity when heated. A common product was a polymer such as poly iso butylene, but there are many other product which do the same job. Lubricating oils contain many other chemicals in their formulation, such as products to reduce the oxidation of the oil when operating at high temperatures, and detergents to keep any solids such as carbon from the burning of the fuel from depositing and causing blockages in the many small oil passages in an engine.

Softening point of bitumen or tar

As these products do not have a sharp melting point like pure compounds, but soften over a range of temperatures, it is not possible to measure the melting point.

Melting point of chemical compounds are usually measured by placing a tiny amount into a glass capillary and heating in glycerine in a flask at a slow rate and noting the temperature when the compound melts.

But tar and bitumen do not melt at a specific temperature, and soften over a temperature range, and this can be used to define tars or bitumens of different types.

The tar and bitumen products have various specifications so that the end user can select the particular product suitable for his application. Softening point range is measured in a similar way to melting point as described above. The product is melted and poured into a small brass ring and allowed to solidify. The ring is placed into a small frame and a steel ball is placed on top of the solid bitumen, and the frame is suspended in glycerine is a round bottom flask. The temperature is raised at a specified rate and the temperature is noted when the bitumen has softened enough for the flat under surface to bulge downwards and also when the ball has sunk completely into the bitumen and has passed downwards through the ring. These two temperatures are called the ring and ball softening range.

Bitumen or tar used in a hot climate will have to have higher softening point than that used in a cold climate. On a hot day the bitumen will tend to soften and the road surface can become sticky if bitumen is too low in softening point.

Penetration

Tar and bitumen and many other products are also classified by a test called penetration. A small metal cup is filled with melted bitumen and brought to a specified temperature where it solidifies. A penetration apparatus is placed over the solidified bitumen with a weighted needle in contact with the surface and held with a locking device. The lock is released for a specified time and under the influence of the weight it

Alan McGown

sinks into the bitumen. The depth of penetration is measured.

Many products have specified penetration tests performed, for example lubrication grease has a specified penetration test to indicate its consistency. The needle of the test used for bitumen is replaced with a cone, and the depth of penetration indicates the consistency.

Chapter 7
Bitumen, emulsions

Key words

*bitumen, coal tar, bitumen emulsion, colloid mill,
thermoplastic, aliphatic*

Bitumen is quite different chemically to tar as
mentioned previously. Tar used to be extensively used
for roads as a by-product of coking coal for the steel
industry. Tarmac at airports was named because tar is
preferred to bitumen for the paved areas around the
terminals and workshops. The reason for this is that
bitumen is easily dissolved by the aliphatic (non
aromatic) nature of both the fuel oils and the bitumen
itself. Aromatic tar is less susceptible to damage from
spilled fuel. In chemistry there is a true saying that like
dissolves like.

Bitumen is made by heating residual or high molecular
weight oils left over from the distillation of crude oil and
blowing air into it. This causes a reaction with oxygen
and polymerisation of organic molecules into higher
molecular weight molecules which produces bitumen,
which like tar is a similar viscous sticky liquid when hot
and which will solidify when cooled (thermoplastic).

Tar is mostly composed of aromatic compounds
(based on the benzene molecule) while bitumen is

composed mostly of aliphatic (straight and branched chain) molecules.

Bitumen is used as a binder when mixed hot with the aggregates used for forming the road pavement surface. Hot-mix is a well known product consisting of liquid bitumen and an aggregate mixture. It is produced in a plant and the product is kept hot until laid out on the road base and compacted with a roller, whereupon the bitumen cools and becomes hard, binding the whole lot together into a smooth hard wearing surface.

Bitumen can be made into an emulsion so that it can be used to make road surfaces without heating. An emulsion is where melted bitumen is mixed under high shear in a colloid mill into water which contains a surfactant (soap). High shear simply means that an extreme amount of force (shear) is applied to the 2 liquids and that breaks up the liquids into small globules.

The globules of bitumen are very small and the surfactant stabilises the mixture so that the globules remain separate. The liquid emulsion when cold is of much lower viscosity than melted bitumen and can even be diluted with water. This is called an oil in water emulsion as the oil phase (bitumen) is in the water which is called the continuous phase.

When milk is homogenised the very same process is used, and because the globules are very tiny they are quite stable and in the case of milk, this avoids the cream coming to the top which used to be common in non-homogenised milk.

There are many examples of emulsions such as milk, paints and cream and many cosmetics. When the cold bitumen emulsion is sprayed onto the base course of a road, the emulsion "breaks", that is the bitumen globules join together to form the sticky mass known as bitumen which then sticks to the stones and binds the whole lot together to make a road pavement just as if you had used hot melted bitumen. The big advantage of the bitumen emulsion is that it can be stored and used cold, whereas straight bitumen needs to be heated to above its melting point to be able to be sprayed onto the roadway.

Alan McGown

Chapter 8
Aluminium refining or smelting

Key words

Hall–Héroult, cryolite, aluminium smelting, alumina, graphite, coal tar pitch

Aluminium is made by the Hall–Héroult process, when aluminium oxide is melted in a "pot" the cathode, which is a bath shaped vessel contained made of carbon, and an electric current is passed through the molten product. The other electrode, called the anode, is a block of carbon attached to a rod and is lowered into the molten aluminium oxide. The molten material in the bath also contains cryolite, Na_3AlF_6 and aluminium fluoride, AlF_3 to reduce the melting point of the mix. Aluminium oxide melts at over 2000 °C while this mixture melts at around 1000 °C.

The aluminium oxide, (Al_2O_3, alumina), is made in a separate factory in the Bayer process where bauxite which is a mixture of aluminium and iron oxides is dissolved in sodium hydroxide and the alumina is separated by crystallisation. The alumina crystals are separated from the liquid and dried at high temperature called calcining.

When the electrical current is flowing between the anode and cathode, the liquid aluminium accumulates at the anodes then falls to the bottom of the pot and is periodically piped away to be cast into moulds. Aluminium smelters contain many rows of such cells.

The reaction is:

$2Al_2O_3$ + 3C + electrical energy $\rightarrow$ 4Al + $3CO_2$

This can also be written in a different type of reaction:

$$Al^{3+} + 3e^- \text{ gives } Al$$

What is happening is that the carbon under high temperature and the supply of electrons from the negative electrode (anode) produces liquid aluminium metal. The electric current causes the separation of the oxygen from the aluminium oxide and this is given off as carbon dioxide gas leaving aluminium metal. This is similar to iron smelting where carbon at high temperature extracts the oxygen from the iron oxide leaving iron.

The voltage is quite low, only about 3 to 5 volts, but the direct current can be thousands of amperes, and thus the aluminium industry is a very large consumer of electricity. Both the pot (cathode) and the carbon electrode (anode) are made up of carbon, which are made by mixing lumps of anthracite carbon with a hot molten binder such pitch. As mentioned before, pitch is made up of very high molecular weight hydrocarbons, which means that the molecule have a very high ratio of carbon to hydrogen.

As mentioned before, pitch is aromatic in nature and the hexagonal benzene rings are almost identical in structure to graphite. Let's say you have a molecule in the pitch consisting of say 7 benzene rings joined together. The 32 carbon atoms each of molecular weight 12, have a combined molecular weight of 384 and the 18 hydrogen atoms of molecular weight 1 have a combined molecular weight of 18. Thus 384/(384+18) = 95.5 % carbon.

The larger the molecule the higher proportion of carbon there will be.

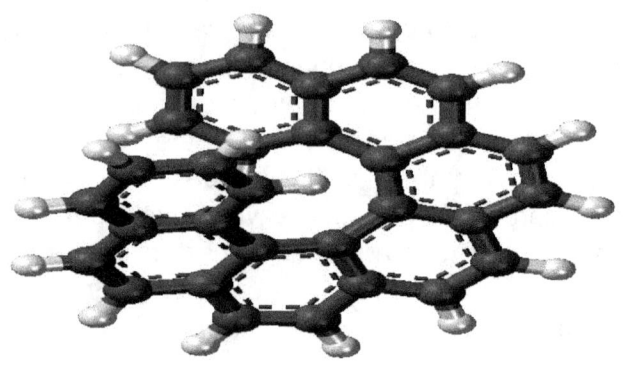

3D Diagram of a high molecular weight aromatic compound. In pitch there will be many thousands of different but similar aromatic compounds like this. The black balls represent carbon atoms and the light coloured balls represent hydrogen atoms.

It does not take much energy to drive off the small amount of hydrogen and you are left with nearly pure graphite, in theory. In practice this is nearly correct, but there are a large number of different aromatic compounds present in pitch and they are not all so regular as the suggested typical molecule. So what really results is an amorphous mass of nearly pure carbon in the form of many tiny hexagonal flakes of graphite joined together at all odd angles.

Back to making the electrodes. After mixing the pitch with pure carbon lumps at a high temperature to liquefy the pitch, the pitch coats the carbon lumps and act as a binder, gluing all the lumps together. The mixture is in the form of a paste and is pressed into blocks under pressure to make "green" electrodes. These are slowly baked and the volatile molecules are distilled out of the block. Further heating causes the hydrogen molecules to leave until only the carbon is left, so that the binder has now turned into carbon and glues the whole lot into solid mass of nearly pure carbon.

The properties of the pitch binder are very important, because the carbon resulting from the pitch can be a little more reactive when being used in the smelting process than the carbon from the carbon lumps. Remember that the pitch which has turned into carbon was applied as a coating to the original lumps of carbon. If the carbon from the pitch reacts a little more than the other carbon, then the binder carbon disappears and lumps of carbon can fall out, commonly called erosion. What this means is that the electrodes which are expected to be consumed in the reaction are consumed at a rate much faster than expected.

This obviously make costs higher than they should be and much research is done in looking for the ideal properties of electrode binder pitch so that it reacts at the same rate as the other carbon in the electrodes.

Alan McGown

My job was to operate a laboratory aluminium reduction cell where we would make electrodes from a constant carbon type but to vary the binder pitch by using pitch from different sources. The physical and chemical properties of the different pitches were measured and recorded and many runs of the cell, each taking one day each were done. The amount of electricity passed through the electrode was measured and the start and finish weight of each electrode was recorded and its actual and theoretical erosion rates were calculated. After many hundreds of runs the properties of the different pitches were graphed against the erosion rate, looking for which properties showed some correlation with higher or lower erosion rate of the electrodes.

The initial graphs showed many seemingly unrelated dots on the sheet. Each different pitch used had a different aromatics content and it could be seen that for each pitch there were many different erosion results – not what we were hoping for. We expected to see groups of results for each pitch type.

I tried a bit of basic statistics based on the assumption that for each pitch type there could be experimental errors and that it was just as likely to produce a high result or a low result. So I removed the highest and lowest result for each pitch type and suddenly the graph almost made sense. Then I made an average calculation of erosion rate for each pitch type and the graphed results produced points on the graph which at last made some sense although not what we expected.

In the end we discovered that the gross aromatics content of each pitch as measured by nuclear magnetic resonance spectrometry, when compared to the erosion rates of each electrode showed that the erosion rate was at a minimum when the aromatics content was around 65%. Coal tar pitch has an aromatics content of around 90 to 95% and the petroleum based pitch we were making happened to have aromatics content of 60 to 70%. Naturally the management were elated because this showed that the type of pitch we were making appeared give a reduced erosion rate as compared to the traditionally used coal tar pitch.

This appeared to favour the petroleum pitch we were making and it ended up being a good selling point. But of course the customers, the aluminium smelters would not implicitly believe the result favouring our pitch, and several trials had to be undertaken to prove that our pitch would be suitable for the anode making on a full industrial scale.

Pitch prills

We started to sell large amounts of our petroleum pitch to the aluminium smelters, but had to find a convenient physical form for the product, so that we to be able to efficiently deliver the product to the aluminium smelters. The pitch could be solidified by cooling and then crushed into a form which could be handled as a bulk solid product. We opted to make prills of the pitch, which is little rounded lumps of solid pitch about the size of match heads. It took a bit of development to find the best way to make these prills.

Alan McGown

The method which was found to work was to have a tub of about 200 litres of hot molten pitch with about 100 holes about 3 mm in diameter in the bottom. This tub was mounted on top of a hollow vertical pipe about 1 metre in diameter and about 20 metres high called the prilling tower. In each hole was a tapered pin and all 100 pins were attached to a plate which would be slowly lowered and raised. When each pin was withdrawn from its hole a small stream of pitch would run off the end of each pin, turn into rounded droplets and fall vertically down the tower. These falling rounded droplets would solidify in the stream of cool air being drawn upwards, and to make sure that they were cooled and solidified before reaching the bottom, they would fall into a stream of water. A conveyor made of mesh would separate the prills from the water, excess moisture would be blown off the prills on the conveyor and take the prills from the bottom of the tower to a nearby stockpile, ready for despatch in bulk containers.

Chapter 9
Chemicals in foods – triglycerides

Key words

triglyceride, saponification, saturated, polyunsaturated, monounsaturated, ruminant, hydrogenation, fatty acid

For a time I was employed as a chemist at an Agricultural College where Dairy and Food Technology and Agriculture were taught. There are many chemical types in foods as normal constituents, and many exist in food whether from plants or animals.

One of the interesting groups are the fats and oils. In this context I am talking about organic oils from animal or plants and not about mineral oils which have been discussed previously.

Fats and oils are really the same thing depending on the melting point, and there are many different types although all have the same basic structure. They are all called triglycerides which is a glycerol molecule with three fatty acids attached. Glycerol is trihydroxy propane

$$
\begin{array}{ccc}
H & H & H \\
| & | & | \\
H-C- & C- & C-H \\
| & | & | \\
O & O & O \\
| & | & | \\
H & H & H \\
\end{array}
$$

Or written differently

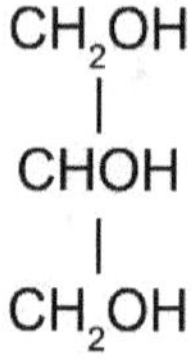

$$CH_2OH$$
$$|$$
$$CHOH$$
$$|$$
$$CH_2OH$$

Fatty acids are a straight chain molecule of carbon atoms with attached hydrogen atoms and at one end is an acid group:

Acid group

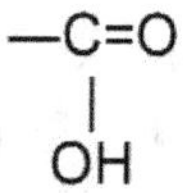

$$—C=O$$
$$|$$
$$OH$$

For example stearic acid is a C18 chain with a carboxyl (acid) group at one end.

$$CH_3- (CH_2)_{16} -COOH$$

A typical triglyceride molecule looks like:

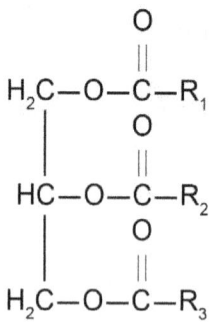

Each of the 3 hydroxyl groups (OH) in the glycerine can react with an organic acid and it is normal for there to be 3 different fatty acids attached to a glycerine molecule, shown as R1, R2 and R3 in formula above.

Triglycerides with high molecular weight fatty acids which are saturated like stearic acid C18, tend to set solid at room temperature, and these are mostly animal fats like butter fat and animal body fat, but palm oil or coconut oil are almost the same composition and contain palmitic acid (C16 chain) as well as small amounts of C 14 and C18 fatty acids.

Saturated fatty acids have no double bonds in the chain, such as stearic acid, palmitic acid. Unsaturated fatty acids have 1 or two double bonds in the molecule for example:

Oleic acid   C18 chain with 1 double bond

Linoleic acid   C18 chain with 2 double bonds

Linolenic acid   C18 chain with 3 double bonds

Triglycerides with unsaturated fatty acids (C18 and one double bond) such as oleic acid tend to be liquids at room temperature and these are almost always from vegetable oils such as olive oil or canola oil.

Triglycerides with poly unsaturated fatty like linoleic or linolenic acid are also liquid and come from linseed oil, safflower oil or sunflower oil and many other plant seed oils.

Triglycerides from animals such as butter fat from milk or fat within meat are usually saturated containing such as stearic and those which come from plants can also contain stearic and palmitic acids. In particular palm oil and coconut oil are primarily triglycerides of stearic and palmitic acids, saturated and very similar to animal fats.

The triglyceride molecule can be broken by a process called saponification which uses concentrated sodium hydroxide at high temperature. This produces glycerine and the sodium salt of the fatty acids which is called soap. Soap was the first surfactant as due to its polar nature with the hydrophilic (water loving) sodium at one end and the hydrophobic (oil loving) C18 hydrocarbon chain at the other end. Remember Palmolive soap; what do you think was the source of oil to make this? Palm oil and olive oil.

To test such fats and oils a gas chromatograph is used to determine the percentage of the mix of fatty acids in the triglyceride. But first the fatty acids have to be separated from the glycerol by the use of sodium hydroxide. The fatty acids in that state are not volatile enough to be injected into the gas chromatograph but if they are reacted with methanol with a boron trifluoride catalyst, the methyl esters of the fatty acids are produced and these are volatile enough to be tested in a gas chromatograph.

An interesting fact is that esters of short chain molecules are sweet and fruity smelling chemicals and most of the sweet odours of fruit are from esters like ethyl acetate, methyl butyrate.

Polyunsaturated foods. In the 1950s and 60s it was discovered that vegetable oils containing triglycerides based on the mono and poly unsaturated fatty acids were thought to be good for preventing the increase of cholesterol in human blood, which appeared to be linked to heart disease, while triglycerides based on the saturated fatty acids appeared to contribute to the increase in blood cholesterol.

The Commonwealth Scientific Industrial Research Organisation (CSIRO) developed a method of treating stockfeed so that ruminant animals (dairy and beef cattle) could be made to produce polyunsaturated triglycerides in their body fat instead of the usual saturated triglycerides.

It was planned that from these animals fed this special stockfeed, meat and dairy products could be produced which would be good for the health of people consuming these.

In cattle fed normal polyunsaturated triglycerides, the rumen (the first stomach) of the animal hydrogenates the triglycerides, they turn into saturated triglycerides and the supposed benefit is lost. The CSIRO method involved protecting the unsaturated triglycerides so that they would pass unchanged through the rumen and the animal would then digest them and the animal fat would then become unsaturated.

This was achieved by taking sunflower seeds which contains mostly polyunsaturated triglycerides and milling them with additional protein into a paste. This was then treated with formaldehyde which crosslinks the protein and makes it into a form which is not digested in the rumen.

Alan McGown

My job was to test the stockfeed and check that it contained the required amount of unsaturated triglycerides. In addition the body fat of the selected animals was check by taking a small sample from each animal and testing it by gas chromatography as described above. It took some time for the animals' unsaturated level to change from saturated to unsaturated but eventually it was achieved. The beef cattle were slaughtered for their meat, the dairy cattle were milked in the usual way and it was shown that indeed the fats in the meat and dairy products had changed from saturated to unsaturated.

All seemed to be going well and commercial outlets were developed for the unsaturated meat and dairy products, but problems started to appear. The unsaturated meat and dairy products seemed to develop off odours (rancidity) and that was eventually shown to be due to oxidation of the unsaturated fats by oxygen in the air. In the end that problem could not be overcome and the whole project was discontinued.

Chapter 10
Fluorine based chemicals

Key words

fluorspar, hydrogen fluoride, hydrofluoric acid, fluorocarbon, hydrofluorocarbon, refrigeration, refrigerant, heat pump, Montreal Protocol, ozone layer, ammonia

Fluorocarbons (CFS's) are a very useful group of chemicals. Hydrogen fluoride (HF) is made by reacting fluorspar, (calcium fluoride, $Ca F_2$) with super concentrated sulphuric acid at 350°C in a rotating steel kiln heated by gas.

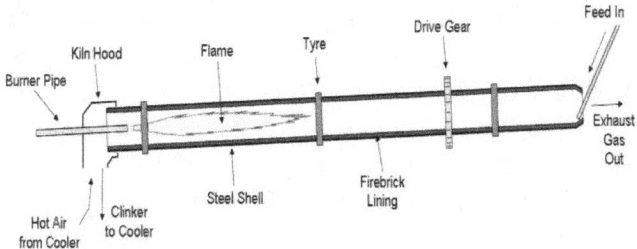

Diagram of a rotary kiln. This one is actually a cement kiln, but a rotary kiln for making HF is just the same. Fluorspar and sulphuric acid is fed into the top of the kiln, HF distils out from the top, and calcium sulphate falls out the bottom. Rotary kilns have many uses for example lime and cement manufacture.

Alan McGown

The hydrogen fluoride distils as a gas out of the mixture and is condensed into a liquid under pressure. What remains is calcium sulphate or chemical gypsum. This had a few uses as a replacement for natural gypsum in glass and cement manufacture and in remediation of sodic clay soils.

The reaction is

$$CaF_2 + H_2SO_4 \rightarrow 2HF + CaSO_4$$

Calcium fluoride plus sulphuric acid gives hydrogen fluoride and calcium sulphate.

Note that this is a good example of a balanced chemical reaction with the number of atoms of each type being the same on the right hand and left hand of the arrow in the reaction.

A group of fluorocarbons are made by reacting carbon tetrachloride, $C\,Cl_4$, with anhydrous hydrogen fluoride in the presence of a catalyst, and then they are separated by distillation. The chlorine atoms knocked off the carbon tetrachloride molecule are combined with the hydrogen from the hydrogen fluoride to make hydrochloric acid, another useful by-product industrial chemical.

The proportion of the different reactions products could be changed by varying the reaction conditions. If the reaction was done at lower temperatures then a higher proportion of the higher boiling point products such as F11 and F12 would be produced.

For example

$$4CCl_4 + 10\,HF \rightarrow CCl_3F + CCl_2F_2 + CClF_3 + CF_4 + 10HCl$$

The fluorine atom replaces 1, 2. 3 or 4 of the chlorine atoms of the carbon tetrachloride molecule, CCl_4 to give a mixture of 4 gases:

1. Fluorocarbon 11 or F11 which is fluoro trichloro methane. CCl_3F, a liquid at normal atmospheric pressure with a boiling point of 23.8°C

2. Fluorocarbon 12 or F12 which is difluoro dichloro methane. CCl_2F_2, a gas at normal atmospheric pressure with a boiling point -29.8°C

3. Fluorocarbon 13 or F13 which is trifluoro chloro methane. $CClF_3$, a gas at normal atmospheric pressure with a boiling point of -81.4°C

4. Fluorocarbon 14 or F14 which is tetra fluoro methane. CF_4, a gas at normal atmospheric pressure with a boiling point of -182°°C

Note that the more Fluorine in these molecules the lower is the boiling point and also the higher the pressure of the liquefied gas.

There were many manufacturers and the products were given a variety of names, for example Freon 11, Forane 11, Genetron 11, F11, R11, P11, CFC11 etc.

CFCs 11 and 12 were the most useful of the group, and were used extensively as refrigerant gases and as aerosol propellants. They readily change from liquid to gas as in aerosol propellants or from gas to liquid and back when used as refrigerant gases. The other useful properties are that they are very stable and have little or no toxicity. In fact F11 continued to be used until recently used as a propellant in asthma spray aerosol products.

Another group called hydro fluorocarbons based on the chloroform molecule, $CH\ Cl_3$. There were mostly used as refrigerant gases and the most important one was R22, difluoro chloro methane. It had a very useful range of properties for refrigeration and there were and are many other types available, all with various useful properties.

Hydrochloric acid made as a by-product of fluorocarbon manufacture is an important industrial chemical. It is also deliberately produced by combining chlorine and hydrogen. Salt is electrolysed by passing a current through salt water, and the products are hydrogen chlorine gas and sodium hydroxide, another important industrial chemical.

$$2NaCl\ +\ 2H_2O\ +\ \Delta\ \rightarrow\ Cl_2\ +\ H_2\ +\ 2NaOH$$

The chlorine so produced is reacted with hydrogen gas to make hydrogen chloride which when dissolved in water is hydrochloric acid.

$$H_2\ +\ Cl_2\ \rightarrow\ 2HCl$$

One of the primary uses of hydrochloric acid is to pickle steel prior to other processes such as painting or galvanising. Pickling means to dissolve the surface layer of iron oxides which are produced when steel is processed by hot rolling. When the iron oxide is removed from the surface it leaves a clean pure steel surface which then forms a surface alloy with zinc when the steel articles are dipped into molten zinc in the process called galvanising. See a more detailed description later.

The company I worked for had more hydrochloric acid than it could sell and other uses were developed. One of these was to react it with limestone, ($CaCO3$) to make a solution of calcium chloride.

$$CaCO_3 + HCl \rightarrow Ca\,Cl_2 + CO_2 + H_20$$

In those days calcium chloride was used to accelerate the curing of concrete during winter months. It had an unfortunate side effect in that if the concrete was not perfectly made and was slightly porous the extra chloride ions existing in the concrete could accelerate corrosion of the steel reinforcements, and after several years, the use of calcium chloride was banned for use in concrete.

Other uses were found which depended on the hygroscopic properties of the calcium chloride solution. Hygroscopic means that the liquid does not dry by evaporation, but actually absorbs more atmospheric moisture. This was used to good effect when mixed with sodium chlorate to be used as a defoliant on cotton crops to remove the leaves on commercial plantations of cotton just before harvest. If sodium chlorate was used as a solution in water to be sprayed onto the cotton leaves it would eventually dry, but the addition of calcium chloride made the solution remain wet on the cotton leaves and become much more effective.

Alan McGown

Refrigeration and air conditioning

These mostly use fluorocarbons and hydrofluorocarbons as they are readily converted from liquid to gas and back, which is what the refrigerant cycle depends on.

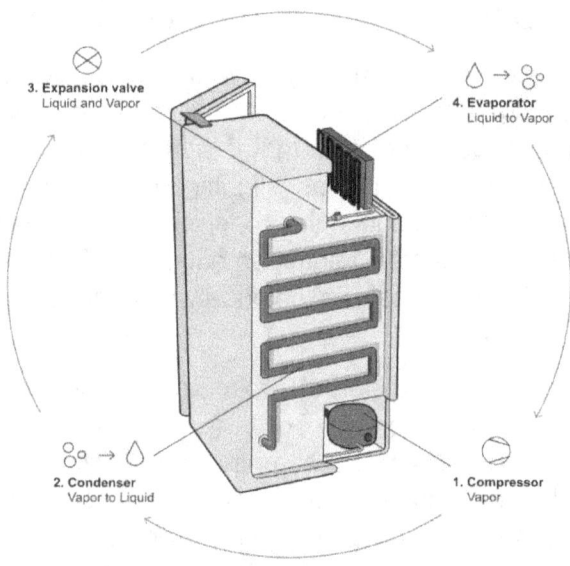

Diagram of a refrigeration cycle. 1. compressor, 2. condenser, 3. expansion valve 4. evaporator

The principle is that when the refrigerant gas is compressed it gets hot and when cooled it will turn into a liquid. When that liquefied gas is allowed to expand from a liquid to a gas it gets very cold.

The gas is compressed and get hot, and then the hot gas is passed through a series of coils or a radiator, called the condenser, and that is cooled by a fan which transfers the heat to the air as it is blown over the surfaces of the pipes. As the temperature of the compressed gas is lowered it condenses into a liquid and flows out of the condenser.

The next stage is the expansion valve which acts as a restriction on the pipe coming out of the condenser and this restriction keeps the pressure on the liquefied gas until it passes through the valve.

On the downstream side of the expansion valve the pressure is lowered and the liquefied gas flows into a similar coil called the evaporator where it evaporates into a gas and the evaporation causes a chilling effect. This is the same chilling effect you feel when you have water on your skin and a breeze blows over it. The evaporation absorbs the latent heat of vapourisation as it turns into a gas. This absorption of heat can be quite large. When you dry your hands under a heated hand dryer at first it seems like the air is not hot, but what is happening is that the water evaporating from your hands is absorbing the heat from the hot air. It is only when your hands are almost dry that you start to feel the temperature of the hot air.

The pump of the refrigeration unit is sucking the gas from the evaporator and compressing it as it returns the gas to the condenser where the cycle starts over again.

So we have the condenser which is hot and the evaporator which is cold and the amount of heat being absorbed by the evaporator is the same amount of heat being given off by the condenser.

Alan McGown

Temperature is measured in degrees for example, Celsius or Fahrenheit. Heat is measured in Calories or these days Joules or kilojoules.

Now back to the refrigeration cycle. As described the condenser is hot and is giving out heat and the evaporator is cold and is absorbing heat. In a refrigerator the evaporator is the cold part and the condenser which is giving out heat is usually at the rear of the refrigerator.

In an air conditioner when cooling indoors, a fan blows air over the coils of the evaporator and the air becomes cool as it gives up its heat to the cold evaporator. The heat absorbed at this point goes into evaporating the gas which passes out side where it is compressed in the condenser and it also has a fan blowing air over the hot condenser. Thus the heat is absorbed into the refrigerant gas indoors and transferred to the air outside where it is wasted.

A reverse cycle air conditioner allows the evaporator and condenser to be interchanged. Then the condenser giving out heat is inside and the evaporator which is absorbing heat is situated outdoors.

Now a really interesting point about air conditioners is that they can appear to be more than 100% efficient which sounds too good to be true. A normal electric radiator is usually close to 100% efficient, that is if such a radiator consumes one kilowatt of electricity it gives out almost the same amount of heat if expressed in kilowatts. An un-ducted gas heater is almost the same but a ducted gas heater may lose half of its heat out put in the hot gas going up the flue. Similarly a wood burning heater may lose 60% of its heat up the flue and be only 40 % efficient.

Air conditioners are called heat pumps, that is they are transferring heat to indoors from outside or the reverse. The air conditioner takes heat from the outside air, thus cooling it and transfers the heat to the air indoors thus heating it. In winter time you will find that the outside unit of the air conditioner is producing cool air. In summer this is reversed and it produces hot air outside and that heat is what it has extracted from the air indoors.

The amount of heat transferred can be measured in kilowatts. An air conditioner may be able to produce say 8 kilowatts of heat but the power to operate the compressor and 2 fans may only consume 5 kilowatts. Thus it appears to be 8/5 =1.6 or 160% efficient.

Aerosol sprays

Another major use of fluorocarbons was as the propellant in aerosol sprays. As described above under refrigeration these gases may exist as liquids when compressed. A 50:50 mixture of R11 and R12 had just the right properties and was widely used until fluorocarbons were banned in aerosol sprays in 1987 after the Montreal Protocol. These days other gases are such as butane, pentane and dimethyl ether are used as the propellant in aerosols.

The way the aerosol sprays work is interesting. At the top of the can is a valve which has a tube (dip tube) connected to it which takes liquid from the bottom of the can. In the formulation inside the can the active ingredients may be dissolved in a solvent such as alcohol or a petroleum derivative, and these are chosen to be miscible (soluble) with the propellant. This means that there is only one liquid phase in the can and what comes up the dip tube is active ingredient, solvent and propellant.

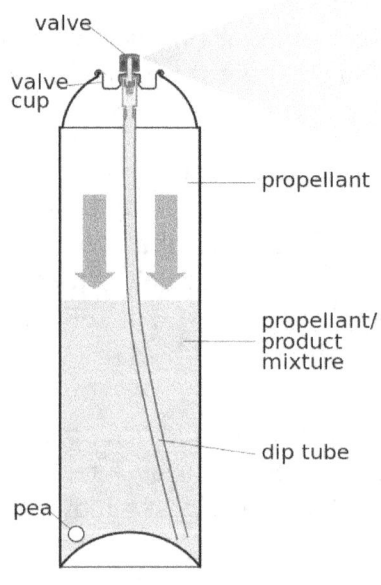

Diagram of an aerosol spray can.

If a space spray such as a flying insect insecticide has been formulated, then there is a large proportion of propellant in the can, possibly up to 40% by weight and the outlet of the valve is simply a hole. When the mixture exits the hole in the valve the droplets of liquid are no longer under pressure in the can and the propellant instantly turns from a liquid to a gas and a very fine, so called aerosol atomised spray is produced.

If a wet spray such as a spray paint has been formulated then a small amount of a propellant is used. Because there is not much propellant in the liquid coming out of the valve it will not atomise as it exits the valve and a mechanical dispersion type of valve is used. You may notice that it does not spray a long distance like the flying insect killer and it is a wet spray, just what is required for paint.

Every different aerosol formulation has quite specific properties and the different properties are designed by choosing the correct type and quantity of propellant, type of valve and solvent in addition to the active ingredients.

Montreal Protocol and banning of fluorocarbons

In 1973 Frank Roland and Mario Molina (research chemists in USA, later awarded the Nobel Prize for chemistry) postulated that because fluorocarbons were very stable molecules they would end up high in the stratosphere where they may be broken down by ultra violet light to give chlorine free radicals which would in turn react with ozone. It was suggested that the ozone layer would be reduced and as this provides life on earth protection against ultra violet light, mankind would all suffer the consequences.

A scientific debate raged throughout the 1970s and 1980s and in 1987 the Montreal Protocol resulted in governments all around the world banning the manufacture and use of these fluorocarbon products. This is a huge topic, there are hundreds of scientific papers and many books on the topic.

Aerosol sprays were developed to use alternative propellants, in each case a liquefied gas such as dimethyl ether and butane.

Alternative fluorinated refrigerants gases were also developed which would break down in the atmosphere before they could reach the stratosphere and thus not contribute to the ozone depletion caused by the previous refrigerants.

Another common refrigerant gas is ammonia with a boiling point of -33°C. Is main disadvantage is that it is very pungent and very alkaline if mixed with water. Note its structure NH_3, and it readily dissolves in water to make ammonia solution NH_4OH a very alkaline material.

Chapter 11
Gypsum and plaster

Key words

gypsum, Plaster of Paris, plaster, porcelain, sodic clay, precipitate

Gypsum is a simple mineral found in deposits where salt has formed in salt lakes. Because gypsum is less soluble in water than is salt, the gypsum precipitates before the salt lakes dry up.

Gypsum is calcium sulphate dihydrate, $CaSO_4\ 2H_2O$. The $2H_2O$ is called water of crystallisation and occurs in many chemicals. The water can be easily be driven off with a little heat at about $150^{\circ}C$

This was done extensively in Paris and became known as Plaster of Paris. What is interesting is that plaster still has some water of crystallisation left and its formula is actually $CaSO_4\ \frac{1}{2}\ H_2O$. When this is mixed with the right amount of water, after a short time it will set hard into plaster and the formula will revert to $CaSO_4\ 2H_2O$.

Plaster board used as wall boards in housing is simply two layers of cardboard separated by about 8mm thickness of set plaster.

If the gypsum has been overheated all of the water is driven off and it loses the magical property of being able to set in the usual way and it is then called "dead burned gypsum".

When plaster of paris is mixed with the right amount of water it turns out to be very viscous (thick) and it has a flash set (very fast set rate) which can be a problem in making plasterboard. Some extra water is added to make the viscosity and set rate more controllable, but that extra water is not used up in making the plaster of $CaSO_4\ 2H_2O$. The extra water remains in the plaster, creates porosity and the plaster ends up weaker in compressive strength than it would be if it had the correct composition.

If that extra water is dried out gently then the pores create a strong suction and affinity for water. This is used to advantage in making plaster moulds for porcelain ware, such as plates, toilets, tiles and hand basins. When the clay for the porcelain article is mixed into a smooth slurry and is poured into a mould made of plaster, the water is sucked out of the clay suspension into the plaster and the clay is then stiff enough to be able to be handled and removed from the mould. Then the clay parts are sent on a conveyor through a kiln which dries out the clay and make the items rigid. It is more than just drying out the clay although that is the first step.

Various temperatures are used depending on the service requirement of the articles. A secondary glaze is applied and the articles are again sent to the kiln where the glaze melts and forms a shiny surface on the items, now known as porcelain.

At the beginning of chapter 10, it was mentioned in the manufacture of hydrogen fluoride, that a by-product called calcium sulphate was produced. As the temperature was about 350 $^\circ$C it was "dead burned" and that calcium sulphate could not be used as plaster. The set was much slower than plaster and this turned out to be an advantage. However we discovered that it could be made to set faster with the addition of a little catalyst called potassium sulphate. While the set rate remained slow, the amount of water which needed to be added was quite close to that required to fully take part in the reaction, and so no water was left over and the mass was quite non porous and very strong.

We found that this type of plaster could be used in some cases as a replacement for concrete, but commercially it was just not viable, as concrete was firmly established and engineers knew its properties well.

The mineral gypsum or its close relative dolomite has long been used to treat clay soils to solve problems with water. When a farmer has paddocks which contain sodic clay the drainage is very poor and crops may not establish and water may lay around. When gypsum or dolomite is applied, the calcium ions exchange for the sodium ions on the clay particles. The net effect is that the clay becomes more like soil, and in the paddocks the drainage magically improves.

Clay can exist in 2 forms, dispersed and flocculated. Sodic clay has sodium ions adsorbed onto the clay particles, the net charges are balanced and each clay particle can exist alone and separate from all the surrounding clay particles. This is called sodic clay and it has most of the properties we associate with clay. When there is not much water present it will be a thick puggy mass which is quite plastic, that is it will deform under force. When a small lump of the clay is added to clean water it will disperse and form a halo around the lump.

The other form is when the sodic clay has had the sodium ions replaced by calcium ions. This makes the clay particles unbalanced in charge and they clump together to be flocculated particles. When a lump of this sort of clay is added to clean water it does not form the halo. This is a simple test to see whether the clay is sodic or not. Only if it is sodic will it respond to treatment with gypsum.

Chapter 12
Fire fighting gases (Halons 1211 and 1301)

Key words
halon, 1301, BTM, 1211 BCF

These two gases are related to the fluorocarbon gases used in refrigeration and aerosol propellants. The different amounts of chlorine and fluorine in 1301 compared to 1211 are there to give the desired boiling points.

The two main ones are:

Bromo chloro difluoro methane (BCF) or halon 1211

This was in common use in yellow hand held extinguishers but was eventually banned as it is slightly toxic. Its boiling point is -3.7 °C which means that it can be expelled from a fire extinguisher as a liquid and it will change to a gas where it is most effective in fire extinguishing. It was developed during World War 2 as a fire extinguisher in tanks, ships engine rooms and aircraft engines.

Alan McGown

Bromo trifluoro methane (BTM or halon 1301)

Halon 1301 is widely used in high value computer enclosures and aircraft engines. Boiling point is -57.7°C which means that it is a compressed gas and is injected into the space where an unwanted fire is burning.

Both of these Halon gases work by the bromine molecule interfering in the chain reaction of flammable materials reacting with oxygen. In high value computer rooms the extinguishing system was set to provide an atmosphere with 6% by volume of Halon 1301. This was sufficient to extinguish any fire but was completely non toxic, so it could be deployed while people were still in the room, but because of toxic products of combustion generated by the fire before the Halon 1301 was deployed, people were usually evacuated. Obviously water sprinkler fire extinguishing systems are not appropriate where high value electronic equipment is being used.

Many aircraft engines are protected by Halon 1301 systems as the engines are usually enclosed. The fire fighting property needs to work very rapidly as there is considerable flow of air through the engine compartment.

In the 1990s the manufacture of these two halons was banned as they have significant potential to deplete ozone and add to global warming. Many pre-existing systems are still installed and there are a few other potential halons available.

Another fire fighting gas, carbon dioxide still used in hand held fire extinguishers, works by simply excluding oxygen from the burning matter. Obviously in an enclosed space there would not be sufficient oxygen to support life, and one had to be very mindful of that if intending to use a carbon dioxide extinguisher.

Alan McGown

Chapter 13
Paper making chemicals

Key words

paper pulp, fourdrinier, drainage aids, flocculants, alum, polyacrylamide,

Paper making is an interesting process where wood chips are made into a slurry of cellulose fibres in water by shredding vigorously in large machines. If the fibres come directly from wood chips, then various chemicals such as sodium hydroxide and sulphides are used to help separate the cellulose fibres from the lignins, which are resins binding the cellulose fibres together in the wood.

There are many different processes of making pulp from wood depending on the type of paper to be produced. Most cardboard is made from recycled paper where the waste paper and cardboard is shredded in large machines and made into a slurry with water. As waste paper contains a lot of dirt, ink plastics and metals it has to be cleaned while in the slurry state before it can be made into paper.

There is no clear dividing line between paper and cardboard, also called paperboard. White paper for printers and photocopiers is almost entirely made from new wood pulp so that it can form a clean white paper without the contaminant usually found in pulp made from waste. You can buy white paper made with some recycled waste and if you look closely you can see small specks of contaminants. Paperboard, for example for your cereal packets is a multi layer laminate with recycled cardboard at the back and white cardboard on the front to allow coloured printing to be produced on the front.

The pulp, whether new pulp or recycled, is mixed with water and is refined by passing through a machine which breaks up all of the pulp into individual fibres suspended in the water and is turned into a slurry. If it is recycled paper then there are a number of cleaning processes to get rid of some of the contaminants. If a lot of the pulp is from newsprint it contains a lot of black ink and must be de-inked before further processing.

The slurry may contain only about 1 to 2% of wood fibres and is quite liquid. This slurry is pumped through a horizontal slot in the headbox onto a moving "wire" screen. The wire is like flyscreens you may have on your windows at home, but are usually finer in mesh and can be up to 5 metres or more in width and may be up to 100 metres in length formed into a continuous loop and is moving at a speed of 3-8 metres per second. The water is drained through the screen and the removal of water is aided by large suction devices underneath the moving wire. Many layers of paper pulp on the wire may be built up in this way depending on the type of paper or cardboard being produced.

The moving wire screen next is pressed between felts which again suck more water out of the pulp. Next the paper still on the wire is transferred to a different wire and is passed over steam heated drums to further dry the paper. At each drying stage the paper pulp becomes stronger until it is strong enough to be rolled up at the end of the paper machine into large rolls which may be up to 6 metres in width and up to 3 metres in diameter.

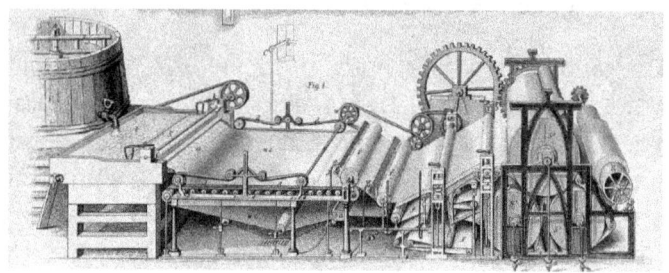

Drawing of an older style paper machine. The headbox is on the left and the fibres slurry is falling onto the moving wire screen. The next section near the large wheel is where the paper still supported on the wire is being pressed against felts. On the right hand side are three large steam heated drums which finally dry the paper before it is rolled up, Modern paper machines are based on the same principles shown here.

If you hold up a sheet of paper to the light you will see that it is not totally homogeneous, that is there are higher and lesser density areas within the paper and this is called formation. High quality papers are required to have a fine formation, but newsprint, the cheapest grade of paper usually has poor formation. Poor formation results in a paper or cardboard with a low strength. You will notice how easy it is to tear newspaper compared to photocopy paper.

You will also notice how newspaper will tear more easily in one direction and that is called the "machine direction", because as the wire screen is moving a high speed when the slurry is applied the fibres tend to line up in that direction and there are fewer fibres in what is called the "cross direction".

Paper formation is influenced by the addition of chemicals called flocculants, which are added to make the water drain more quickly from the pulp on the wire screen. Unfortunately the more chemicals added makes the formation worse and there is a fine balance between adding sufficient quantity and type of flocculant to achieve rapid water drainage whilst still giving acceptable formation. Some flocculants mostly made of poly acrylamide can improve the dewatering of the pulp on the wire whilst minimising poor formation.

Before synthetic polymer flocculants became available most paper and cardboard was made using aluminium sulphate (alum) to flocculate the fibres. If the pulp as a slurry contained only long fibres then there would be less need of a flocculant as the fibres would bridge over the holes in the wire screen. But the slurry also contains a lot of short fibres, the main action of a flocculant is to make the short fibres into clumps of fibre which would not pass through the wire and will be retained in the forming paper.

Alum is also acidic and as it is made by reacting aluminium hydroxide with sulphuric acid.

$$2Al(OH)_3 + 3 H_2 SO_4 \rightarrow Al_2 (SO_4)_3 6 H_2O$$

Alan McGown

It can be a very crude process and the alum is usually supplied as a solution in bulk tanker loads from the chemical plant to the paper mill. Alum has many industrial uses such as flocculating contaminants in drinking water and clarifying water in farm dams, sewage treatment works or swimming pools. The reaction is the same as with the fine fibre in the paper slurry. Basically the aluminium hydroxide precipitates with the suspended contaminants or short fibres and makes them clump together into a floc, in the case of drinking water this allows contaminating clay particles to settle more rapidly that they would otherwise.

$$Al_2 (SO_4)_3\, 6\, H_2O + fibres \rightarrow fibre\ floc\ 2Al(OH)_2 + 3\ H_2 SO_4$$

You can see that this reaction is just the reverse of the above reaction when the alum was made. The result is that the sulphuric acid is then liberated into the water draining from the paper pulp or the water draining from the flocculated clay and the solution is then acidic. Thus the waste water from paper mills tends to be acidic and causes problems at the sewerage works as the sulphate ion is acted upon by sulphate reducing bacteria causing the liberation of hydrogen sulphide or "rotten egg gas".

Paper made by this process also contain small amount of acid and this causes such papers to be acidic and it tends to fall apart after many years. Valuable documents can have this acidity neutralised by saturating with a gaseous alkali such as ammonia, but this is a very expensive process as it involves placing books and documents into a pressure vessel when the pressure is initially lowered with vacuum pumps to remove most of the air from the pores of the paper, and then the ammonia or other gaseous alkali is pumped into the tank to saturate the pores of the papers to react with the sulphuric acid and neutralise the paper.

Acid free paper can be made by using alum free systems, and it is not as cheap and easy as using alum, but is becoming more popular for high quality printing paper, mostly because of the cost of removing sulphuric acid from the paper mill waste water, rather than any desire of the paper companies to produce acid free paper.

Technology is advancing continuously with new chemical advances, better instrumentation to better control the process and larger and faster paper machines.

Starch

You may be surprised to find that considerable quantities of starch are used to make paper. Wheat starch, rice starch and potato starch are all used.

Starch as it comes from the grains is a very linear structure with very long molecules like in the diagram below, but consisting of thousands of glucose molecules. When suspended in water it exists as tiny particles which if you could look at them appear to be tight little bundles of chains and the suspension is of low viscosity.

When the starch in water is cooked, the long molecules unravel to make long chains which makes a highly viscous solution which is commonly applied to the dry paper as it comes from the paper making machine. When the concentration of starch in water is high it may even form a semi-solid gel.

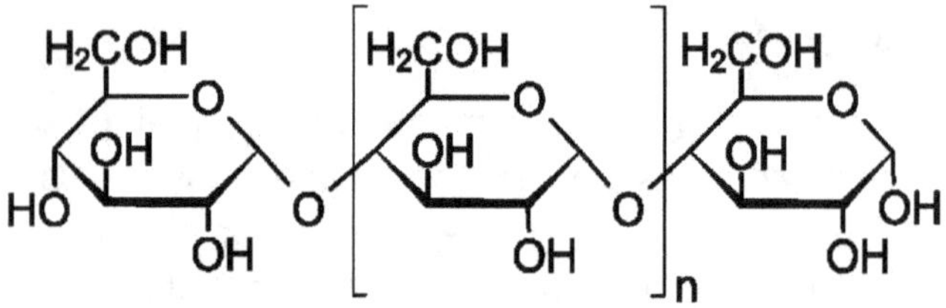

Structure of amylopectin or starch. Note that these are not benzene rings as each has an oxygen atom in the ring. These rings are actually glucose molecules. This is only a partial structure as a molecule may have many times more glucose units as can be shown here.

To visualise this phenomenon, imagine a large bucket of rolled up chains, each ball the size of roadway gravel, which are not attached to each other, and can easily be poured from the bucket, almost like a liquid. When the starch suspension is cooked, the rolled up chains unroll to become long chains which get tangled up with each other and the whole lot then becomes a very viscous liquid. This is how starch works in cooking, thickening foods like porridge or making gravy.

The starch when applied to the surface of the dry paper and dried acts like a film of resin and makes the paper stronger, more water repellent and smoother, which helps when the paper is to be printed.

Chapter 14
A simple but dangerous chemical –
Carbon disulphide – CS$_2$

Key words

carbon disulphide, sulphur

Carbon disulphide is used as an intermediate in manufacture of pesticides and rayon which contain carbon and sulphur. Carbon disulphide is one of the most flammable materials and will ignite from a hot water bath. It is also used in laboratories as a solvent with particular properties, but one which needed to be used with extreme caution. It is a dense liquid of boiling point 46°C, burning with a typical sulphur blue flame and giving off sulphur dioxide and carbon dioxide.

These days CS$_2$ is made by reacting sulphur in gas over catalysts. But up until the 1970's it was made by a very crude and what appears to be very dangerous process. A sealed retort was filled with lumps of carbon which was then ignited until the whole mass of carbon was glowing. The vapour space was purged with steam to drive out the air and then molten sulphur was poured into the top of the retort.

The following reaction took place:

$$C + 2S \rightarrow CS_2$$

Alan McGown

As the glowing carbon had a temperature of around 1600°C the carbon disulphide was distilled out of the retort and cooled in a condenser to below its boiling point to become a liquid.

Chapter 15
Fibres, composites (fibreglass), ballistics

Key words

glass fibre, carbon fibre, para-aramid fibre, FRP. GRP, CRP, epoxy, nylon, polyester, polypropylene, cotton, cotton ginning, linen, rayon, silk, weaving, knitting modulus of elasticity, Youngs modulus, e-modulus, spiral winding, pultrusion, thermoset resins

Fibres are used to make fabrics for clothing and furnishings, and also for industrial uses which are mainly composites (fibreglass) and also ropes and cords.

Clothing and furnishing

The main fibres used are cotton, wool, nylon, polyester, rayon, flax and sometimes mixes. All these are made into a yarn by various means and the yarn is then woven or knitted into fabric. Cotton, flax and wool have been used over the centuries, but chemically they are very different. Cotton is almost pure cellulose and is made into yarn after the cotton bolls are removed from the cotton plant. The yarn consist of almost pure cellulose which is just a polymer of glucose. See the section on starch in chapter 13 which is also a polymer of glucose.

Alan McGown

A cotton yarn is just all of the fibres from the cotton boll aligned and twisted together. The twist is needed to keep all of the fibres together and to give the yarn some strength. Strength of cotton yarn is increased by blending in polyester yarns which makes minimal difference to the comfortable feel of cotton garments.

Linen is a little similar to cotton with a large proportion of cellulose (92%) but as flax is made from the stems of the flax plant it contains other products such as starch, sugars and minor components. Linen is not as popular as it once was as the costs of processing the flax plant into a yarn is higher than other fibres.

Wool as you know comes from sheep, alpacas, goats etc and like silk is almost pure protein. Silk is still a premium fabric due to its attractive properties but as it comes from the cocoons of silk worms, the processing into any quantities of yarns is labour intensive and expensive. Wool is available in large quantities and it takes a lot of mechanisation to turn wool into yarns. Wool is also a premium product but unless it is made from the thinnest fibres (around 18 micro metres thick) it does not have a good skin feel, and not all wool is of this micro fibre type.

Rayon sometimes called viscose, is the first semi-synthetic yarn and was first made in 1892. It is in fact mostly cellulose which comes from wood pulp, by dissolving the purified pulp in sodium hydroxide, treating with carbon disulphide to make a thick syrup which is then extruded into thin filaments and treated with sulphuric acid to regenerate the cellulose. The filaments are then combined into a yarn and twisted to give it some strength. Rayon has excellent skin feel, and has been called artificial silk due to its apparent similarity to silk.

Nylon is one of the very first fully synthetic yarns, which is actually quite similar in chemical structure to wool and silk.

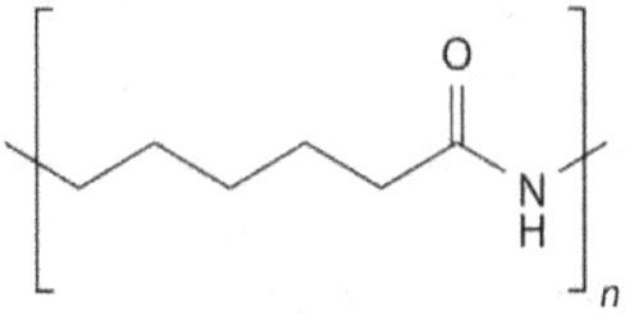

Nylon 6

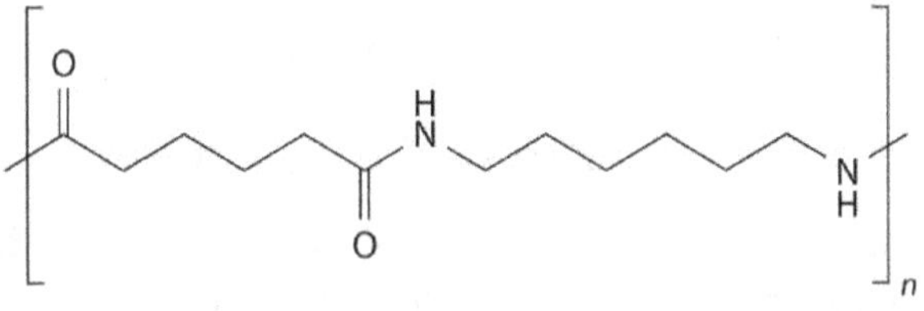

Nylon 6,6

Chemical structure of nylon. Note there are two main types depending on which monomer is the starting material. The properties are very similar.

The first nylon was nylon 66 made from hexamethylene diamine and adipic acid. Both have similar properties and are made by polymerisation of the monomer, then extruding the product into thick filaments about 3 mm diameter which is then chopped into chips about the size of matchheads. These chips are then melted and extruded through very small orifices to make yarns. A single filament called monofilament is used for ropes or fishing lines. All synthetic yarns are first made into chips before spinning into yarns as the chips are a convenient way of transporting the materials prior to their further processing.

Polyester, acrylic and polypropylene yarns are made in the same way as nylon, just different monomers are used. These materials just discussed are all thermoplastic because they melt when heat is applied which allows their extrusion into yarns. Thermoplastics are also important engineering plastics as they can be moulded by injection moulding. See chapter 16.

Industrial applications of fibres

Composite structures (like fibreglass)

Wood is the original natural composite material, being made of cellulose fibres, and it is lignin which is the glue which holds the fibres together. Wood is still used extensively in amateur built aircraft and boats. In wood directly from the tree, all the fibres are aligned and most of the strength is in the longitudinal direction. Wood is easy to split along the grain and that can be a disadvantage.

Plywood is used which places multiple layers of wood with the fibres in each layer aligned at 90 degrees to the layers on either side, so that it does not suffer from the easy splitting characteristic of wood. Just like in fibreglass structures, the fibre contributes most of the strength and the resin is just there to hold them all together.

In those products commonly called fibreglass or composites or Fibre Reinforced Plastic (FRP) or Carbon Reinforced Plastic (CRP) or Glass Fibre Reinforced Plastic (GRP), the fibre contributes almost the entire physical properties of the product.

The resin is in most cases thermosetting polyester or epoxy resin and has a tensile strength far below the fibres, in fact people in the industry say that the resin is there just to keep the water out when used for making fibreglass boats. This is somewhat correct as the resin is really there just to glue the layers of fibres together.

The design of composite structures is based on placing the fibres into the laminate so that the fibres are under tension, that is so that the fibres will tend to be stretched. The designer simply calculates the expected stresses and specifies enough fibres in the right places to resist the tensile (stretching) forces whilst minimising the amount of stretch.

Well that is what is supposed to happen, but as we saw in the 1995 America's cup race in San Diego, the Australian entry One Australia broke in two. It seems that the designers in their quest to make the boat as light as possible did not design enough carbon fibre in the right places and the stresses involved exceeded the design expectations and the actual tensile strength of the composite structures was exceeded and caused the hull to break. During the race the boat was being bent like a banana due to the tension applied by the back and fore stays (wire ropes) attached to the mast. When the sails are pushed by a gust of wind the whole boat tends to bend in the middle and if the bending force is high then something has to give. Sometimes it is the back stay, sometimes it is the mast, but on this occasion it was the hull which could not resist the bending force.

I should explain tensile strength and tensile modulus. Reinforcing fibres are tested by placing a short strand of fibres into a machine which pulls it apart with an increasing force and the fibre stretches until it breaks. Both the force applied and the amount of stretch are measured continuously and are usually shown on a graph, called the stress/strain curve. Stress is the force and strain is the stretch.

When the force being applied is enough to break the strand, the tensile strength is reached. The distance the strand stretches before it breaks is also measured and that is used to calculate the tensile modulus. Tensile modulus or Young's Modulus or E Modulus is an expression of the amount of stretch in a fibre. A high modulus fibre such as carbon has a much lower stretch than glass fibre. Consider a piece of carbon fibre with high tensile strength and high modulus, side by side with a piece of glass with low tensile strength and low modulus. If both are tested together in the tensile testing machine it is the carbon fibre which will break first.

Because of its relatively low tensile modulus (compared to carbon or aramid fibres) the glass fibre keeps stretching whilst the carbon fibre reaches its tensile strength limit. Or to use an example more people would be familiar with is to use a piece of string side by side with an elastic band. Let's make the string of a higher tensile strength that the elastic.

Tensile testing machine. The jaws holding the specimen move slowly apart, the stress (force) and the strain (stretch) are measured and recorded on a graph.

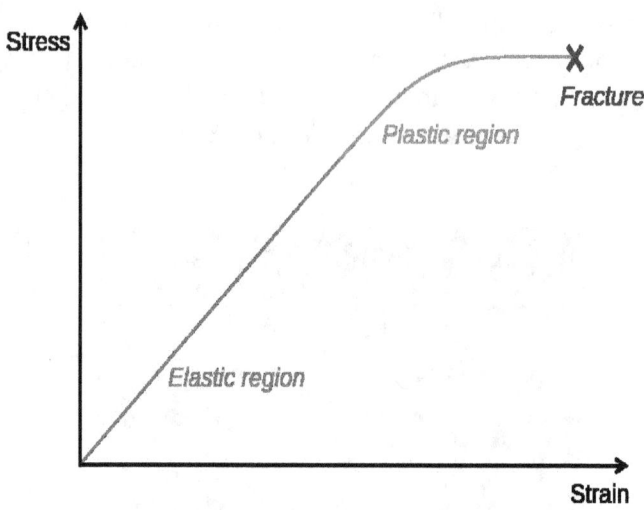

Stress/strain graph from tensile testing machine. As the stress (force) increases so does the strain (stretch). Initially the specimen is in the elastic range, and there the stretching is reversible. When it reaches the plastic region it starts stretching without much increase in the stress. Metals particularly have a very large plastic region where they will deform and that is not reversible. Most fibres have a very small plastic region and fracture soon after. Strong materials will have a steep line and low strength materials a line at a lower angle.

When the load increases both stretch a bit until the limit of stretch for the string is reached. The string breaks when the load reaches the tensile strength of the string, but obviously the elastic does not break as it has not reached its limit of stretch. The instrument registers that the load has suddenly decreased but the distance between the clamps holding the string and elastic stays the same and is increasing. Eventually after a long stretch the elastic reaches its limit of stretch and its tensile strength and it breaks. The interesting point is that the stronger one has broken first because of its low stretch or higher tensile modulus.

It is a popular misconception that carbon fibre or aramid fibre composites are much stronger that glass reinforced plastics. It is commonly said that carbon fibre has a higher strength to weight ratio than glass fibre, which is slightly correct as the carbon fibre is significantly lighter than glass fibre. On the matter of strength, composite products using fibre reinforcement are not designed for strength, as it is that the actual strength should never be approached by the product in normal use. Composite products are designed for stiffness and that takes into account not the tensile strength of the reinforcing carbon fibre which is only a little more than the tensile strength of glass fibre, but the tensile modulus which is 3 times higher in carbon fibre than in glass fibre.

Consider the spar in a glider wing which is in the shape of a steel I-beam. That means if you look at the end it is in the shape of the capital letter I. This spar takes almost all of the stresses in flying and the wing panels which are bonded to the spar caps transfer the lifting forces to the spar. It has an upper and lower spar cap made up of multiple layers of carbon fibre fabric embedded in epoxy resin. In that structure the majority of the fibres are longitudinal as that is the direction where the stress will be felt as the aircraft wing flexes to take the load. Thus the two spar caps are designed to operate in tension when loaded. The upper spar cap is in tension when the aircraft is sitting on the ground and the wings droop down. The lower spar cap is in tension when the aircraft is flying and the wings are supporting the whole weight of the aircraft and thus they tend to bend upwards.

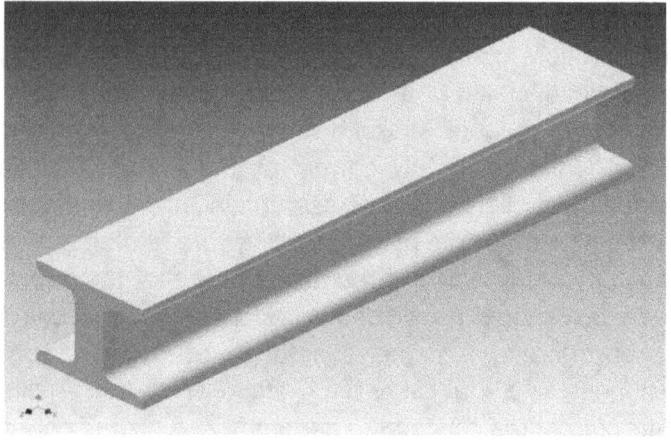

Diagram of an I beam. This is the usual shape of the spar in an aircraft wing.

The two spar caps are bonded to and separated by a web which is frequently made of fibreglass, and the small stresses on the web are mainly compressive. In a thick wing the tension forces on the spar caps and the compression forces on the web are not as severe as in a thin wing. So in a low performance training glider with short thick wings, the stresses on the spar are small compared to a high performance state of the art glider with long thin wings. And thus in a low performance training glider there is little point in the extra expense of using carbon fibre in the spar caps, as fibreglass is perfectly capable of resisting the tensile forces in the spar caps.

Now we come to the exact reason why carbon fibre is used. As the tensile forces in the spar caps are high because of the thin wings of a high performance glider, the spar caps will tend to stretch and allow the wings to bend. Glass can be used but as it has a lower tensile modulus(high stretch) it will need more glass fibre in them in order to limit this stretch and that will make the wings much heavier than if high modulus(low stretch) carbon fibre was used.

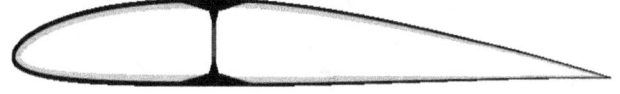

Diagram of a wing in end elevation view showing the spar shaped like an I beam.

A modern high performance glider aircraft is designed with only just sufficient carbon fibre in the spar caps to limit wing bending. So the fibre strength does not come into it, except in extraordinary circumstances of exceeding the allowable wing loading in a severe dive and pullout. It is all about putting enough fibre in the place where the tensile forces will be, to make a structure stiff enough for the intended service.

The same goes for boat hulls where "top hat" shaped stiffeners are used to make the hull stiff enough. The top hat stiffener acts just like the spar in an aircraft wing – the forces from the hull panels are transferred to the stiffener which resists stretching or bending.

Alan McGown

Carbon fibre

Carbon fibre is used extensively today to make high tech products, basically as a replacement for fibreglass, but as it has a higher tensile modulus than glass fibres parts made from it may be lighter and/or stiffer than they would be using glass fibres.

Carbon fibre is made in a similar way to the making of pitch previously described. A polyacrylonitrile polymer is made into fibre by extruding the thermoplastic, and then heated in a nitrogen atmosphere to drive off the hydrogen and nitrogen molecules leaving a skeleton of graphite which is cooled and rolled up as a yarn. This yarn is then woven into various fabrics to be used for reinforcements in the composites industry.

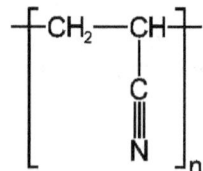

Acrylonitrile monomer structure. Thousands of these monomers are joined end to end, and once the high temperature has driven off the hydrogen and nitrogen you are left with chains of carbon atoms in the form of carbon fibres

Para-aramid fibres

With trade names like Kevlar, Twaron and Technora, para-aramid fibres are not used extensively in composite application as there are two problems. Whilst they are lighter than carbon fibre and have higher tensile modulus than glass fibre they do not perform well in compression (such as in the upper spar cap of a wing when it is flying) and their toughness. It was thought that the lightness and high tensile modulus would make it a good reinforcement in wing panels and boat hulls and that is true. But if repairs ever have to be done, the toughness of these para-aramid fibres means that when the surface is being prepared for painting by sanding, the fibres fluff up and resist being sanded.

Meta-aramid fibres

The most common meta-aramid fibre is known as DuPont Nomex and is extensively used when woven into a cloth to make flame retardant firemen's or car racing driver's clothing. Before Nomex became available wool was the most flame retardant fibre and also cotton treated with Proban, a commercial treatment containing phosphorous compounds.

Flame retardancy does not mean fireproof, as the only fibre that is flame proof is asbestos, and due to its carcinogenic (cancer causing) properties it obviously cannot be used. Flame retardant fabrics simply will not burn when the source of ignition is removed. In fact the test involves igniting a strip of fabric with a Bunsen burner and measuring the time for it to self extinguish.

Glass fibres

Glass fibres are used in composites in several main forms:

Alan McGown

Chopped strand mat

As the name suggests, chopped pieces of glass fibre about 30 to 60 mm in length are laid out and glued together with a plastic compatible with the polyester resin which is common in the composites industry. This product gives the lowest cost and lowest performance composite product, as the fibre are all in random directions. But the one advantage chopped strand mat has it that once the resin has been applied it is deformable in 2 directions whereas a woven fabric will not.

Chopper sprayed fibres

A continuous fibre glass roving, which is a continuous bundle of glass fibres together, is fed to the chopper gun, where during operation it is chopped into short lengths about 5-20 mm. A pressurised resin and catalyst is also fed to the chopper gun where it is mixed with the chopped glass and the whole mass of resin and chopped glass fibres is sprayed out onto a mould. The ratio of glass to resin is controlled by the operator and the hardener for the resin is also fed to the chopper gun but usually in a fixed ratio to the resin.

Before the spraying of the resin and glass fibre takes place, a wax based release agent is applied to the inside of the mould so that the finished product will not stick to the mould. The next step is to apply a gel coat. This is basically a polyester resin paint which is sprayed into the mould before the mass of resin and glass fibre and will form the outside of the product.

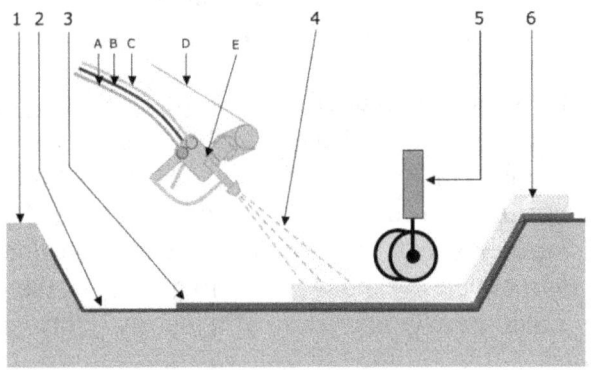

Diagram of fibreglass construction using a chopper gun. This example is of an article where the outside of this article is what will be seen, for example the hull of a boat. 1. The mould usually also made of fibreglass 2. release agent 3. gel coat 4. spray of mixed resin, catalyst and chopped glass fibres 5. roller applied to the surface by hand to consolidate the resin and glass layer 6.resin and glass layer A. liquid resin pressurised B. compressed air to power the chopper and spray the mixed resin and glass C. pressurised catalyst D. glass roving supplied from a bobbin E. chopper gun

This is a very effective way of applying multiple layers of resin and glass to the mould. The mould has a very smooth surface on the side where the fibreglass is to be applied. Imagine making a small boat hull where the mould is boat shaped but the inside of the mould has the shape of the outside of the boat hull. When the layers of resin and glass fibres are spray applied to the inside of the mould they will set after the desired time and a boat hull can then be lifted out of the mould.

Finishing then consists of grinding off rough edges and fitting out other components inside the hull.

The main two disadvantages of using chopped fibres or chopped strand mat are that the fibres are laid in random direction and large thicknesses are required to achieve enough stiffness for the product. If the glass/resin layer is thin the structure will be floppy. Because a high thickness is required for stiffness it is also quite heavy and has a relatively high resin content.

In a well designed composite product the reinforcing fibres are ideally aligned in the direction where the stresses are expected to be and stiffness is achieved by the use of webs and bulkheads and stiffening panels.

On the matter of stiffness a commonly used rule-of-thumb is that if you double the depth of a beam you quadruple the stiffness. So making stiffer panels means that you simply make them thicker. Thicker and stiffer panels can be made using a sandwich construction where a lightweight material such as a stiff foam plastic or honeycomb is used between layers of composite. All the stresses are in the outer layers and the foam or honeycomb is in what is called the neutral layer which experiences very little stress.

This is just like the spar of an aircraft described before, where the two spar caps take all of the bending stresses of the wing and the web in the middle is just there to keep the spar caps apart.

Woven fabric

A higher quality composite product can be made using a woven glass or carbon fibre as the reinforcement. There are many different weave patterns, from a balanced weave where there are equal number of fibres in the warp and weft to ones which have a higher percentage in the warp direction, called unidirectional. Warp is the longitudinal direction of the fabric on the roll, and weft is the cross direction.

Woven fabrics are widely used either as glass or carbon and in a wide range of fabric weights and different weave types. For high performance composites the crimp in the yarns is a limitation. Crimp is where each fibre goes under then over those at 90 degrees and when the fabric in use is stressed this crimp tends to straighten out and the fabric does not achieve the tensile modulus of the fibres used in that fabric. This disadvantage is overcome with the use of multiaxial fabrics.

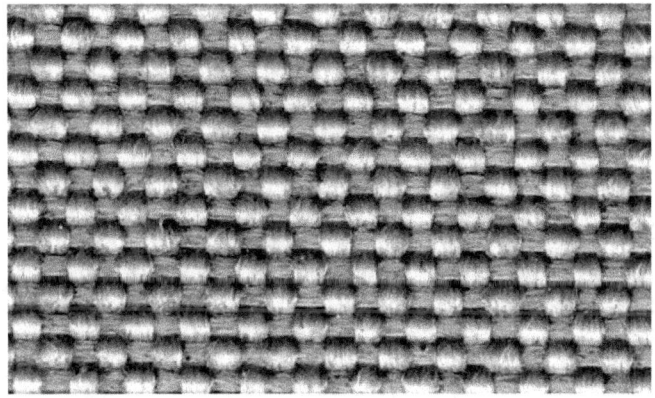

Close up of woven glass fibre.

Alan McGown

A loom weaving glass fibre.

Multiaxial fabrics

These are not woven, and can have fibres all in one direction or with some in plus and minus 45 degrees directions. These are mostly glass fibre fabrics and are formed in a large machine which stitches together layers of glass fibres.

Again either woven or multiaxial fabrics allow the designer to place the fibres where the expected stresses are and in doing so will achieve a lighter and stiffer product than if the cheap, weak and heavy chopped fibre or chopped strand mat is used. When the composite engineer designs a composite construction for a particular application it is mostly unidirectional fibres which will take the tensile loads. Because there is no crimp in those fibres, the tensile modulus inherent in the fibres is achieved in the final composite construction, and the reinforcing fabric is able to perform exactly as designed.

Prepregs

The highest performance composites are made by a process of using a woven fibre (usually carbon fibre) already preimpregnated with resin. This mixture of resin and reinforcing fibres can be made under very closely controlled factory conditions to have the lowest possible resin content and having most of the fibres aligned in the direction of the highest expected stresses, by using closely specified layers of unidirectional and cross directional fibres. The resin in most cases is epoxy resin, which is chosen to have the highest interlaminar shear strength, which means that the resin when cured will resist delamination. Strength failures of composites usually occur by delamination of the reinforcing layers rather than by other failure modes such as tensile strength failure.

The epoxy resin in prepregs is specially chosen to have a slow rate of reaction between the two components, the resin and the hardener. The mixed resin is applied to the selected fibre fabrics is a coating machine and then the layers are combine under moderate pressure to make a prepreg which is a leather-like thick fabric maybe 3 mm thick where the resin is partially cured and the prepreg loses its stickiness. Then a separation plastic film is applied to one or both sides of the prepreg and it is cooled to maybe -10 degrees celsius and rolled up.

The cooling of the prepreg causes the curing reaction of the epoxy resin to virtually cease so that the prepreg whilst kept cold, stays in a soft formable form. When the end user of the prepreg wants to use it to make products, the prepreg is cut into the desired shapes and placed by hand into the mould in one or several layers. Then a plastic bag is placed over the uncured prepreg and sealed around the edges with a tape and then a vacuum is applied to the whole lot. Air pressure then compresses the prepreg into the mould where it very evenly takes up every detail of shape of the mould, and most importantly it compacts the layers tightly together so that adhesion between the layers is maximised.

Then the whole lot is placed into an autoclave which is just a very large oven, while the vacuum is maintained. The high temperature starts the curing reaction of the epoxy resin and the prepreg turns solid and the product then has its final form. The high temperature curing of epoxy resin really enhances the stiffness and strength.

Prepregs are only used where the highest performance products are required such as in the wing parts of commercial aircraft as there is a considerable cost of using this method of manufacture. Part of the cost is the need to keep the unused product refrigerated until it is ready to be used.

Pultrusion

This is a process to make rods of composite used for the inner core (not the light conducting fibres) of optical fibre cables. I shaped beams for non electrical conducting ladders or other innovative products are also made by the pultrusion process.

The reinforcing, usually glass fibres is threaded through a die and resin is applied to the fibres just before entering the die. The die is shaped like the cross section of the article being made. For example if a fibreglass ladder is made the die is shaped like an H and produces long sections which are the side beams of the ladder.

The die is heated and causes the resin to cure rapidly, and it cools a little later to make the completed rod which may be continuous and be rolled up on a large diameter drum or be cut into lengths as it exist the machine. As the glass fibre coated with resin approaches the die it is compressed by the small diameter of the die and any air or excess resin between the fibres is squeezed out and the optimum ratio of glass to resin results.

Spiral winding

Yacht masts and spars, fishing rods, radio antennae, launching tubes for shoulder launched missiles are just a few of the products made by spiral winding. As the name suggests, fibre are wound around a mandrel which is just a long, slightly tapered stainless steel rod. Resin is coated onto the fibres before they get to the mandrel and a high tension is applied which squeezes out any air or excess resin between the fibres. In most cases carbon fibre is used because most products require the lightest weight and the stiffest walls possible. Once the winding is completed the fibre and resin still on the mandrel can be heated or the product can be allowed to cure at room temperature. Once the resin is cured, the mandrel is pulled out to leave a hollow cylindrical product. Post-cure heating is usually used to make the resin come to its strongest state.

Resins used in composites

Thermosetting polyester is by far the most common resin used in composites and it does a job good enough for the applications. There is no point in using a higher performance resin such as epoxy if you are using a cheap reinforcing fibre such as glass. It is all horses-for-courses. However if a high performance product is being designed then both a high performance reinforcement and a high performance resin is used.

What do I mean high performance resin? A polyester resin as I said before is just there to stick together the reinforcing fibres. If a fibreglass structure is stressed by bending beyond its design then usually the resin will pull apart and you will get delamination as the resin will always have a much lower tensile strength than the reinforcement.

Thermoset resins and FRP industry

The resins used in composites or "fibreglass" are all called thermoset resins. Thermoset resins are a viscous liquid and when mixed with a suitable catalyst, turn into solid or cure usually with the evolution of some heat. If these thermoset resins when cured are heated, they do not melt. Compare this with thermoplastic resins like nylon, polyethylene, polypropylene and thermoplastic polyester which turn into a viscous liquid when heated and back to a solid when cooled. I'll talk more about thermoplastic resins and their applications later.

Back to thermoset resins, they give out heat when curing and the curing can be accelerated by heating. This can be a problem resulting in a limited "pot life". When a thermoset resin is mixed with the catalyst it immediately starts to react and cure. When the mixed resin is applied to the fibreglass a lot of the heat is taken by the glass and a reasonably slow cure results, but the mixed resin in the pot does not lose heat to the same extent as that resin applied to the fibreglass and cure reaction is accelerated by it increasing temperature in the pot. So what can happen is that the resin the in pot cures and sets solid before it can be used, but the resin on the fibreglass still has not cured. This is a design problem for the chemist who designs the fibreglass/ resin method.

One easy way of overcoming this problem is, as mentioned before, that the resin, catalyst and fibreglass can be mixed as it is used in the chopper gun, and the resultant mixture is sprayed into the mould. But then you have the problem of a relatively high resin content composite, which makes it heavy, and short fibreglass strands which makes it not very strong. This is OK for the many applications in the composites industry when high strength and low weight are not required.

Alan McGown

Higher performance composite products are usually made with epoxy resins which have a higher tensile strength and modulus than polyester resin, but epoxy resins are more expensive. It always is "horses for courses" in the composites industry, which means that low performance products will be cheap enough to be economic but high performance products will use expensive reinforcing fibres and resins as that is the only way to achieve the required performance where high cost is not a problem. Examples are:

Low performance, heavy, not very strong, cheap – caravans, small boats, furniture, training class gliders etc

High performance, light, stiff, expensive – motor racing cars, Olympic rowing shells, racing yachts, aircraft etc.

Para-aramid fibres

Para-aramid fibres are also used in many technical applications.

Ballistic protection

The most prominent of these applications is in bullet proof vests, military helmets and panels in military vehicles and ships. The original ballistic helmets and vests were made of multiple layers of nylon fabric, but they were much heavier than those made of para-aramid (Kevlar, Twaron or Technora). High modulus polyethene known as Dyneema or Spectra is also used. The fibre is woven into a plain weave fabric and it is used in multiple layers to resist the bullets or fragments.

A bullet proof vest may use up to 24 layers of fabric stitched together into the vest and there are many different specifications depending on the expected threat level. It is no use for police to wear a heavy military vest which will stop high velocity armour piercing projectiles (bullets), when the expected threat level for police is low velocity hand gun lead projectiles.

As you can imagine from this example, there are many possible design structures and the higher velocity hard projectiles will require more layers to protect against than for low velocity soft projectiles.

Basically the higher the expected threat level, the heavier the construction (more fabric layers) is needed to prevent the projectile from penetrating.

The ballistic fabrics work to stop the bullets just like a soccer net stopping the soccer ball. If you see a slow motion picture of a soccer ball hitting the net, you will see a circular wave which moves outwards from the impact site of the ball, and this spreads the energy of the soccer ball over a large area. It is the spreading of the energy over a larger area which takes the kinetic energy from and stops a bullet. Obviously if the bullet has a very high kinetic energy it may still penetrate some of the fabric and more layers need to be used.

In addition, for a ballistic structure to defeat hard high velocity armour piercing projectiles, it needs a hard ceramic panel at the front to break up the projectile, and the para-aramid fabric catches the fragments of projectile and ceramic.

Alan McGown

I worked in assisting customers specify ballistic structures in army helmets, police vests, panels for warships, panels in armoured personnel carriers and panels in VIP vehicles for heads of state.

Tensile application

Optical fibre cables use glass fibres to conduct the light signals over long distances and a cable with glass fibre strands will have many fibres each taking a different signal. When the cables are installed into underground ducts they are pulled through the ducts with a rope. As the lengths of cable are considerable, many kilometres will create a lot of friction and as the glass fibres are relatively weak they may break due to the pulling forces involved.

To overcome this potential breakage of the glass fibres, large amounts of para-aramid fibres are laid into the cable during manufacture, and it is the para-aramid fibres which overcome the pulling forces. The high modulus (low stretch) of the para-aramid fibres is used to limit the stretch of the whole cable and thus the glass fibres are protected from the pulling forces during cable installation.

Carbon fibre which would also be useful in this application due to its even higher tensile modulus, is not used because of its ability to conduct electricity means that lightning strikes would destroy the optical fibre cables.

Friction application

Historically asbestos was used to make brake pads for motor vehicles, but the danger of asbestos short fibres means that it is no longer allowed to be used. Aramid short fibres can be made into pulp which is rather like cotton wool. This aramid pulp is mixed with resins and minerals, where the aramid pulp holds the lot together before it is pressed into the shape of brake pads. Only aramid fibres are tough enough to resist the very high mixing forces involved in this process. These brake pads are baked at high temperatures where the resin is decomposed and the result is a brake pad where the aramid fibre and the minerals create friction when pressed against the smooth brake disc.

Rubber applications

Aramid short fibres are mixed with rubber and then rolled out into a sheet form to make high temperature resistant gaskets. High pressure resistant hoses used in underground mining are made by using aramid fibres spiral wound or braided in the same manner as you can see in your garden hose. Similarly aramid fibres are used in high temperature rubber heater hoses in automotive applications. Also vee belts and toothed belts used for driving alternator, air conditioning and power steering in motor vehicles now use para-aramid fibre as the load bearing member to prevent stretching of the belt

In these application, the para-aramid replaces nylon or polyester reinforcement which has much lower tensile modulus (high stretch) and much lower temperature resistance.

Chapter 16
Plastics

Key words

monomer, polymer thermoplastic, thermoset, polyethylene, polypropylene, polystyrene, nylon, PVC, polyurethane

Thermoplastic polymers

As there are many thermoplastic polymers (plastics), I will only describe only a few of the most commercially important.

Thermoplastic means that the polymer will melt when heated and will form a solid when cooled. This property make thermoplastics suitable for a lot of forming processes, such as injection moulding, see chapter 18.

Polyethylene

Polymer means many units joined together and the simplest and one of the earliest plastics is polyethylene (polythene) which is made of many units of ethylene joined together.

Ethylene gas $CH_2=CH_2$ reacted with a catalyst at high temperature gives a molecule in a chain structure known as polyethylene.

$$CH_3-CH_2-CH_2-CH_2-CH_2-\text{etc etc}$$

The more ethylene units in the molecule, the harder it is and the higher its melting point is. Polyethylene is a milky white plastic with many uses from plastic bags and milk bottles to engineered products made by injection moulding.

One very interesting use is as woven fibres used in bullet proof vests. In normal polyethylene, the molecules are very long and folded and looped like a bucket of rope and this makes it quite stretchy (or low modulus as previously described).

Polyethylene fibres can be stretched at a temperature just a little less than the melting point, and what happens is the molecules straighten out and when cooled they have a high modulus (low stretch). Just like aramid fibre which is used in bullet proof vests because of its high modulus, polyethylene takes a share of that market.

Polyethylene when used in plastic film or rope for outdoors applications is very sensitive to ultraviolet rays (UV) which break the bonds between the molecules. Luckily UV inhibitors can be added to the plastic while is it being formed.

Polypropylene

Chemically this is quite similar to polyethylene but it is harder, higher tensile strength, tougher and has a more shiny surface. It is widely used for food containers and woven bags for wool bales, fertiliser bags, but these outdoor applications need a UV inhibitor just like polyethylene.

Both polyethylene and polypropylene have long straight chain molecules which are very similar to paraffin (petroleum) waxes. One definitive way to identify them is to burn a little and when extinguished, the white smoke smells just like when you extinguish a wax candle.

Polystyrene

Styrene (vinyl benzene) is a benzene molecule with virtually a polyethylene molecule (vinyl) attached and when polymerised it is the vinyl part of the molecule which reacts to form long chains with the benzene molecules sticking out. Polystyrene is hard, water clear brittle plastic which is mixed with plasticisers or other resins to make it more useful. However the main use is the styrene molecule in many other plastics such as thermosetting polyester as previously described in the composites industry and in styrene butadiene rubber for tyres.

Poly vinyl chloride or PVC or vinyl

This is chemically similar to polyethylene except the starting monomer is vinyl chloride, CH_2=CH Cl and its properties are similar. PVC exists in 2 main forms, one is as a rigid thermoplastic used for drainage pipes. It can be brittle owing to it being filled with mineral powder (ground up rocks) to make it rigid and cheap. The type usually called vinyl is a very flexible form used as a film or upholstery or synthetic leather or as a coating on electrical wires. It is flexible because it contains plasticisers which is a type of chemical dissolved in the resin.

Nylon

Nylon was one of the first polymers made in the search for synthetic silk to be used for parachutes, or wool to which it is chemically very similar. Invented by Du Pont, the name nylon is said to come from the New York and London laboratories of Du Pont where the research was being carried out. Nylon usually does not contain plasticiser like PVC does and its properties can be altered by controlling the molecular weight as is done with polyethylene and polypropylene.

Nylon is the name of a group of similar products known as polyamides and is made by reacting 2 chemicals, hexamethylene diamine and adipic acid (each of these has a chain of 6 carbon atoms) which produces nylon 6,6. The other major type is called nylon 6 and is made by polymerising caprolactam which contains the amide group at one end of the molecule and an acid group at the other end. The chemistry is more complicated than I wish to reproduce here, but there is a large amount of information available in the literature for anyone who is interested.

One of its most common uses as fishing line called a monofilament, where its high tensile strength and stretchiness give it the desired properties. The nylon in the form of chips about the size of match heads passes from a hopper into a heated screw where it partially melts. Then it is extruded through little holes in a heated die and the continuous filament is cooled by air and it solidifies and is wound up on spools ready to be used.

The group of products known as nylons are commonly called engineering plastics because they are used to make products which have mechanical end uses and this often involves being formed by injection moulding. See later section on patternmaking, toolmaking moulding and casting.

Thermoset polymers

This is another separate category of plastics and the most commercially significant types are epoxy and polyester which are used primarily in the composites (fibreglass) industry. They turn from viscous liquid to solid by a reaction between the two components resins (epoxy) or by reaction with a catalyst (polyester). The other significant property of thermoset resins is that once reacted and formed into products they do not have the property of thermoplastics. When heated to a high temperature they will decompose rather than melt. Decomposition means that they will start to react with oxygen in the air and usually start to burn.

Polyurethanes

This is a very large group of thermosetting plastics which are used to make many articles, such as rubber type products such as wheels, shoe soles, adhesives and sealants, engineering parts, car parts, paint, foamed plastic for upholstery and block and sheet rubber products.

Foam products as used in upholstery

This is an interesting application as the foam is made in a very large machine in the form of a loaf (like a loaf of bread but about 3 metres wide 2 metres high and as long as necessary). The reaction of the ingredients starts as soon as they are mixed and the mixture flows onto a long conveyor and swells as it progresses along the conveyor and by the time it reaches the end of the conveyor the reaction has completed and a large band saw is cutting it into large blocks ready for further processing.

The two chemical ingredients are a product called polyol and another, toluene diisocyanate which combine to make the rubbery substance. Various aerating agents are also added at the same time, such as water which makes bubbles of carbon dioxide or fluorocarbon 11 or pentane which turn from a liquid to a gas from the heat of reaction. The bubbles of gas may join up and make the plastic as an open cell foam which means that each bubble has burst and all of the calls are interconnected.

Adhesives

Many synthetic plastics and other products are used as adhesives, but I won't go into detail as there are many, many types. One main property of joining articles by adhesives is that the surfaces to be joined must be clean and allow the adhesive to wet the surfaces. Some materials such as polyethylene, polypropylene and Teflon are *almost* impossible to join with adhesives.

Many materials can be joined with adhesives if the surface has been abraded to make it smooth on a macro scale, but rough on a micro scale. For example when joining wood the surfaces are smoothed with

Alan McGown

abrasive paper and an adhesive such as PVA is applied. This partially soaks into the smooth surfaces and once dried, the join can actually be stronger than the wood itself.

Chapter 17
Semi-permeable membranes

Key words

microfiltration, ultrafiltration, nanofiltration, reverse osmosis, desalination, dialysis

Synthetic membranes used in industry should be thought of as a filtering mechanism for ultrafine particles. Ordinary filtration is achieved by use of a filtering medium such as a bed of sand in the case of filtering swimming pool water. The space between the sand particles allows the water to pass through usually under the force of gravity, while the suspended unwanted particles are larger than the spaces and are caught and retained in the sand.

Other types of filtration such as cartridges use either a cloth or paper or foam plastic which similar to sand filtration, catches the suspended particles in the spaces between the fibres. Eventually all of the spaces become blocked with particles and the flow rate decreases. Filters are then cleaned of the particles by backwashing with clean water and this water including the particles is disposed of.

Membranes work in just the same way but the particle size can extend down to even the tiny size of ions in solution. In general, the pore size determines what

145

type of filtration is applicable. The usual sizes of the particles able to be separated by this technique are:

	Particle diameter microns (micrometres)	Particles removed
particle filtration	10 to 1000	sand, dirt
micro filtration	0.1 to 10	yeast
ultra filtration	0.01 to 0.1	bacteria
nano filtration	0.001 to 0.01	viruses
reverse osmosis	less than 0.001	ions such as Na+ Cl-

Membranes are everywhere in nature. The human body uses many membranes to separate different substances. For example the lungs allow gases to pass through the membrane to remove carbon dioxide and to re-oxygenate the blood. The intestines allow soluble nutrients to pass through. The kidneys, skin and all body cells are membranes, each one performing a specific task.

Osmosis is the driving force in many of these separations and the principle is that a concentrated solution on one side of the membrane becomes more dilute as it draws water through the membrane. Reverse osmosis as used in sea water desalination, simply uses pressure on the feed to force water through the membrane leaving the feed solution more concentrated.

Osmosis through membranes is also the driving force in plants transpiration of moisture. For example consider an orange tree and look inside an orange. The small cells containing juice are small sacs of membrane and contain concentrated sugar, acid and salt. The effect of osmosis is to draw water up through the plant so as to dilute the contents of the cells. The same thing happens with the leaves.

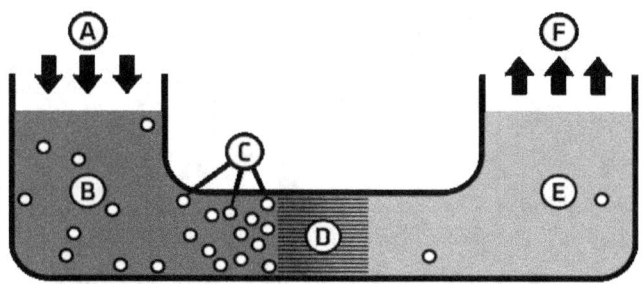

Diagram of the principle of membrane separation. The dilute feed, say seawater, is under pressure and the permeate is quite pure water. A. pressure applied to B. feed solution, C. ions in the feed, D. membrane, E. permeate solution with reduced ions, F. flow of desalinated water

How are synthetic membranes manufactured? In the early days, membranes were just a flat film of plastic usually cellophane, (regenerated cellulose). The amount of liquid which will pass through a membrane is very small and so a large surface area was required. This could be achieved by making the cellophane in a tubular form and in the 1930's laboratory tests showed that proteins from milk could be separated by putting the milk into a tube of cellophane and suspending this in water. Artificial kidneys and dialysis followed in the 1940s.

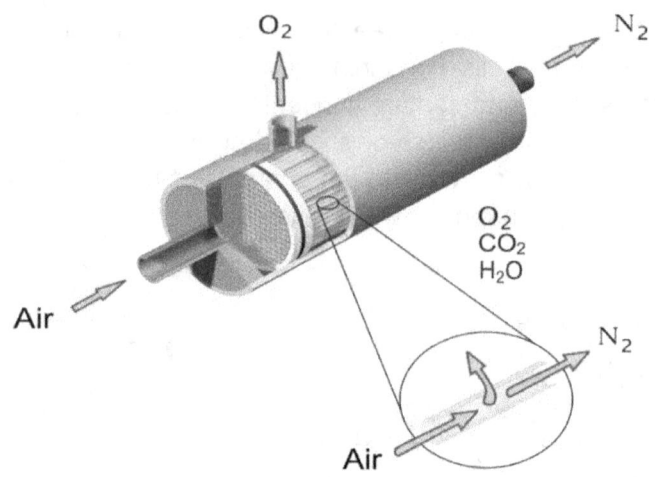

Diagram of how membranes are used. The membranes are tubular and in bundles, and many of these modules are connected to allow a reason able total flow rate. This diagram shows gas separation, but this is the normal arrangement for most membrane applications.

Tubular membranes in the form of straws (see diagram) were developed to allow larger commercial separations to occur, as this structure had not only a large surface area but is also quite robust and able to withstand high pressures without crushing. Commercial salt water desalination typically uses tubular membranes of around 1-2 mm diameter and many metres in length.

Membranes typically use thermoplastic such as nylon or polypropylene extruded in the tubular form. The pores are created by mixing into the molten plastic, a vegetable oil which will be soluble at that temperature, but not soluble at room temperature. Once cooled to room temperature the plastic solidifies and then the oil is washed out of the plastic with a suitable solvent.

At the melt temperature the oil is completely dissolved in the molecular form, and when it is extracted from the solid plastic it leaves behind the tiny pores. When viewed through a microscope the plastic looks like a piece of sponge where the pores are interconnected, and one can visualise it being a filtering medium. Obviously the pore size and total pore volume has to be controlled so that the plastic has the desired microfiltration properties.

Chapter 18
Patternmaking, toolmaking moulding and casting

Key words

pattern, toolmaking, CNC, sand moulding, injection moulding, tool

Patternmaking is not about drawing pretty patterns on paper. A pattern is a model of the article to be made by moulding,

As mentioned before, many articles are made by moulding. Cast iron and other metal goods are moulded by pouring the molten iron or aluminium into a mould made of sand.

The first stage is for the patternmaker to make a pattern, which is simply a model of the article to be made. Patterns are usually made of wood or a suitable plastic. Imagine a garden seat of cast iron is to be made. The pattern is exactly the same dimensions as the intended article.

The mould is made in by placing the pattern in a box then sand is poured into the space around the pattern. The sand contains a thermosetting adhesive and when heated it solidifies and forms a mould in the shape of the pattern. A join is made at the time of making the mould to allow the pattern to be removed. In casting metal parts a hole is made in the top of the mould to allow the molten metal to be poured into the mould. When the metal has cooled and solidified the mould is broken, the cast article is removed from the sand mould, and then the sand is recycled into more moulds. Numerous sand moulds are made from the original pattern.

In casting precious metals a process called the "lost wax method" is used. A pattern in the shape of the intended article is made of wax and the sand mould is cast around it. The whole mould is heated gently and the wax pattern melts and is run out of the mould leaving a cavity in the shape of the pattern. Obviously the pattern is destroyed and so only one article can be made.

Toolmaking for moulding plastic usually uses a metal tool or mould, which is made by milling the cavities into a block of metal with a CNC machine (Computer Numerically Controlled). The CNC machine is a large automated router which means that it has a high speed rotating blade similar to a router used by carpenters to shape the edges of timber. The shape of the cavities required are designed on a computer and once set running the machine mills the required cavities automatically in a block of metal.

Usually multiple cavities in the shape of the article to be made are milled and connected by small passages. Plastics are usually moulded under pressure called injection moulding, where the molten plastic is forced through small passages into the shaped cavities. The metal tool also contains separate passages for cooling

Alan McGown

water or oil to flow through and cool the tool which allows the molten plastic to solidify. The tool usually splits into two parts once the plastic has solidified to allow the now solidified articles to be removed.

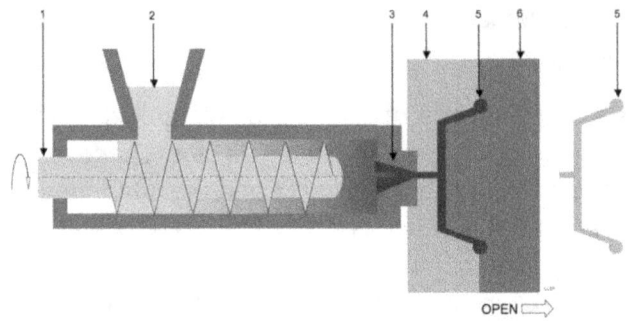

Diagram of injection moulding of plastic. 1. Screw transporting the plastic, 2. hopper containing pelletised plastic, 3. molten plastic, 4, half of split tool or mould, 5, cavity filled with molten plastic and finished article removed after cooling and solidifying.

On all moulded products of plastic glass or metal, if you look closely you will see line or ridge which indicates where the two halves of the mould were joined.

Chapter 19
Galvanising steel

Key words

zinc, galvanising, flux, alloy, sacrificial anode

Galvanising is a process to impart corrosion resistance to fabricated steel products. A zinc coating on the surface acts as a sacrificial anode protecting the steel from corrosion. Steel structural beams which are used outdoors would be subject to corrosion if not protected. Paint can be used but his is subject to damage during construction or aging. A galvanised steel surface is very durable and will not chip off or be damaged by rough handling as would happen with a painted surface. If any damage does occur then the electrolytic protection will still be effective even if damage causes bare steel to exist in small patches and such areas will continue to get long term protection from corrosion.

Galvanising is in basis a simple process, steel articles are dipped into molten zinc.

The surface of the steel needs to be cleaned first in a process called acid pickling. Paint must first be removed by abrasive blasting. The cleaned steel parts are dipped into hydrochloric or sulphuric acid to remove rust, millscale and welding slag. Millscale is just a mixture of iron oxides which form on the surface during the hot extrusion process of forming the steel into beams or pipes. Welding slag is very similar. Then the steel parts are dipped into a flux solution containing zinc ammonium chloride, which allows the molten zinc to wet and react with the steel surface and to form an alloy.

The molten zinc at a temperature of 480 degrees is held in a large tank up to 20 metres in length, 5 metres in depth and 2 metres in width. The steel structures to be galvanised need to be carefully designed with holes drilled to allow the acid, flux and the molten zinc to penetrate all areas of the structure and then be drained out before being lowered into the molten zinc.

The molten zinc actually reacts with the surface of the steel forming an alloy containing both zinc and steel. This alloy is black and is harder than both the zinc and the steel.

In addition, the designer needs to take into account the high temperature involved, as it will cause expansion and contraction of the steel part which can result in the fabricated structure bending or buckling when it cools after being withdrawn from the zinc tank.

The company I was employed by also manufactured galvanised steel drainage grates as well as floor grating used for industrial applications. The drainage grates were manufactured to an Australian standard and there were 6 classes from lightweight grates suitable only for pedestrian traffic up to very heavy grates suitable for airport taxiways. Each different grade of grate has to meet a performance specification which is a maximum deflection (bend) whilst under a certain loading.

I operated a small laboratory which tested a certain number of grates from each production run. The test involved supporting the grate at the edges in the way intended by its design and putting pressure in the centre via a wooden pad of 250 mm square. A large hydraulic ram pressed on the central pad and the pressure was gradually increased up to the specified limit for each grate. Our testing machine could go up to about 30 tonnes. The deflection or bending of the grate was measured up to its maximum rated load and when the pressure was released no residual bend was allowed.

Galvanised electricity tower and roadside railing.
These would have a limited life if not protected
by galvanising.

Alan McGown

The grates were not only supplied for use in roadways, many large projects such as coal loaders, oil refineries and other installation with conveyors, need an extensive network of galvanised steel grating as walkways. Buying these was not a simple matter.

The whole project was designed on a computer using a computer aided drawing program (CAD), which included the walkways with a tolerance of plus or minus 5 mm. and their steel framework supports This means that when the individual grating panels were made in accordance with the CAD engineering drawing, each of the thousands of panels could be no more than plus or minus 5 mm different in actual measurement from what was shown on the CAD drawings. Often there were thousands of identical panels and they had to fit the steel supports which were being supplied by a different manufacturer. In addition many panels were of complex shapes to fit with other parts of the structure such as pipes and ducts.

The CAD engineering drawings were supplied by email and reloaded into our computer. There our draftsmen had to check the details and make drawings of every individual panel, give it a number and make another drawing (by CAD) showing where each panel fitted in the whole structure, so that the customers construction contractors could assemble every one in its correct place, just like a massive jigsaw pattern.

Once the panels were made, the Quality Control (QC) staff had to physically check each numbered panel against the drawings, to be sure that we were supplying exactly what was specified. If for example each panel in a one km long stretch of walkway was the full positive tolerance (5mm) too long then by the time the assemblers got to the other end the panels would be 1.5 metres too long, and that is a massive problem as the supports are built to support each panel in 3 places. So you can see that the QC had to be very strict.

An interesting point about metals is that minor bending can occur without leaving a permanent bend and this occurs in the elastic range of the metal, such as steel springs. Every metal is different and a lot depend on the particular alloy and the hardness or temper of the metal. Refer back to the section on aluminium foil and its hardness. A metal in the hard state is usually more elastic in bending than a metal which has been annealed or made soft by heating.

Once you bend a metal more than its elastic limit, and get into the plastic range, then the molecules actually move in relation to each other and you then get a permanent bend. You can try this with a metal paper clip which has been straightened out. Hold it by one end and you can flex it without leaving a permanent bend, but then bend it over a small radius and you will find that the bend remains.

If bending just short of the elastic limit occurs frequently enough the molecules and or grains can also move in relation to each other and cracking can start. This is the cause of metal fatigue which in aircraft is the reason why a metal aircraft has a finite life and must be scrapped after so many hours of flying and take offs and landings.

It is also interesting that fibre reinforced plastics appear to have no fatigue life and theoretically should last for an infinite number of deflections without deterioration. This was confirmed a few years ago by the Melbourne Institute of Technology which tested a wing from a glider by flexing it for the equivalent of more the 30,000 hours of flying.

Chapter 20
Geotechnical Engineering, soils, rocks, gravel, roads

Key words

roadbase, plasticity, clay, California Bearing Ratio, compaction test, dynamic cone penetrometer

Geotechnical companies test construction materials for properties which are important in the construction industry. For example soil beneath an intended bridge or building needs to be tested for its bearing ability.

A large vehicle mounted drill is sometimes used to take samples of the soil before construction begins. This is then tested for its degree of plasticity which is caused by its clay and moisture content. As was explained in chapter 4, in the section of properties of materials we mentioned plasticity, which is the ability of a material to deform when a load is placed on it.

When a structure is to be built, it is important to know that the soil beneath will support the structure without deforming. The same requirement exists for making a road. The more clay contained in the soil beneath the roadbase dictates that a certain thickness of good quality roadbase gravel needs to be built up. Roadbase gravel is laid down in layers and compacted with a vibrating road roller when it is in a moist condition. As you could imagine, this becomes very dense and hard and if constructed properly will support traffic loads without deforming.

This roadbase being laid down is tested for compaction by lowering a radioactive probe into a small hole in the roadbase and measuring the amount of radiation absorbed by the roadbase. A reading is given in density units. A well compacted roadbase will achieve a density of around 2.4 tonnes per square metre which would then indicate that it is ready for application of the bitumen top surface.

The major cause of roads failing is due to insufficient compaction of the roadbase before the top bitumen layer is applied. The quality of the roadbase is also important here as if it contains too much clay it will never be capable of being compacted up to a high density when laid and compacted.

Roadbase materials are also tested prior to being used in a test called California Bearing Ratio (CBR). Before the road construction is commenced, a sample of roadbase gravel from the supplier is compacted with moisture into a mould 100 mm diameter and 100 mm deep. This is to simulate how it would be compacted during a road construction.

Then the plug of compacted roadbase gravel is taken out of the mould and soaked in water for 4 days, after which it is tested in a machine to determine how much it will deform under an applied load. A good quality roadbase material with a high proportion of gravel will deform little, but a roadbase material with a lot of clay present will deform a lot, just as it would if used to make a roadbase. This test is used to select suitability of proposed roadbase materials. These properties are all specified by Civil Engineers when designing roads and take into account the expected loads of traffic and the bearing capacity of the soil beneath the roadbase. If extreme traffic loads are expected sometimes a concrete layer is used beneath the roadbase.

Army geotechnical engineers testing a gravel airfield in a war zone This test is done to be sure that it has been compacted sufficiently, so that it can bear the weight of large military aircraft without the surface breaking up. The slide hammer is just above the right hand of the standing soldier.

Compacted soils beneath house slabs and earth dam walls are also required to be compacted to provide good and even bearing capacity to ensure that over time it does not slump. A test called dynamic cone

Alan McGown

penetration test (DCP) is done. For example house foundations or earth fill dam walls are tested to check that the compaction has been performed correctly by the earthworks contractor.

The apparatus called a Dynamic Cone Penetrometer, is a cone of about 25 mm diameter connected to a steel rod of about 20 mm diameter. This has graduations every 100 mm in length. A steel drop hammer is used to force the cone and rod down into the compacted soil, and the number of blows per 100 mm is an indication of the degree of compaction. A large number of hammer blows per 100 mm indicates a well compacted soil and a small number of blows indicates a poorly compacted soil. The values are specified by the engineer who designs the house foundations or dam walls. The soil can be tested down to a depth of several metres if necessary. If the values specified by the design engineer are not reached the earthworks contractor may be called in to repeat the compaction.

Chapter 21
NATA accredited laboratories

Key words

NATA, proficiency testing

These days many companies employ laboratories to conduct independent testing on their products, and they usually require the laboratory to have accreditation by an independent body to certify that the laboratory is capable of providing accurate, non biased testing in the particular field.

NATA accreditation means that the laboratory has been examined by the organisation called National Association of Testing Authorities (NATA) and has been found to be performing tests according to specific standards, and the accuracy and precision has been verified by proficiency testing. There are other organisations which may carry out the accreditation.

The following is a small sample of the types of laboratories I have experienced which are required to have NATA accreditation.

Coal testing

To certify to the buyers of coal (i.e. power stations, steel makers) that it meets the required specification and that the buyer will pay an agreed price

Alan McGown

Concrete testing

To certify to the buyers of concrete that it meets the required specification, and that the construction company has confidence in the future integrity of the structure.

Pathology testing

To ensure that tests are carried out to allow doctors to confirm diagnoses on patients without mistakes which could be life threatening.

Geotechnical testing

To confirm the physical properties of soils under proposed structures like bridges, buildings and roads.

Before such laboratory accreditation was common, many products were supplied which were sub standard or not quite suitable for their intended purpose. For example concrete was supplied to projects like bridges, buildings and other structures, not conforming to the specification by the designing engineer. In these cases the concrete may deteriorated to such an extent some years later, that the structures had to be demolished, and naturally the owners were not impressed. These days most concrete being supplied is tested to Australian or other Standards.

To achieve accreditation from an organisation such as NATA, the laboratory has to:

- show which testing standards are being used relative to the field in which they are operating

- show results of proficiency testing in comparison with other laboratories operating in the same field. This means to on a continuing bases, perform tests on a "blind sample" and report the test results to the proficiency testing company

- show example of sample handing processes which meet strict requirements

- show examples of documentation and reporting which meet strict requirements

- submit to annual audits of all of the above to maintain their accreditation.

To have proficiency testing carried out, an organisation which arranges proficiency testing, sends a sample to many laboratories operating in the same field. These laboratories test it and return the test results to the proficiency testing company. Statistical analysis of all results shows the average results and how each laboratory compares with the other laboratories. There are very strict rules about how far the test results can deviate from the average. Any large deviation from the overall average is queried and the laboratory must determine the reason for the deviation, and must show that they have taken corrective action.

Alan McGown

Chapter 22
Concrete and cement

Key words

cement, klinker, concrete cancer, spalling, cement water ratio, slump test, Portland cement

Many people say cement when they are really talking about concrete. There is no such thing as a cement road or a "seement pond." Cement is the reactive ingredient in concrete and the other materials are sand and gravel.

Cement called Portland cement, is made in a large rotary kiln as shown in chapter 10. The main raw materials, clay or shale and limestone are fed into the rotary kiln at about 1450 degrees, heated by coal where the ash of the coal ends up being incorporated into the cement. Cement clinker in lumps comes out the bottom end and this is chemically the same as cement, and is next ground into a fine powder which we know now as cement

Portland cement reacts with water and the surface of sand in the concrete mixture to form a hard layer. This layer also is formed on the surface of the gravel in the mixture. The amount of these 3 main ingredients is based on their particle size. The fine gravel is there to fill the spaces between the stones or large gravel, the sand and cement is there to fill the smaller spaces between the small gravel.

In this way a high density mixture is formed which will be quite waterproof with no spaces left between the particles. If the composition is not correct, there will be minute spaces or porosity which will allow water and air to penetrate with disastrous consequents which I will describe soon. Concrete strength is enhanced by a higher cement content as long as the correct amount of water is used. Many concrete laying contractors like to add more water to the mixed concrete to make their job of laying the concrete easier. This practice severely reduces the strength and increases the porosity of concrete.

Excess mixing water damages concrete because only an exact amount of water is needed for a complete reaction with the cement and that amount of water will be combined into a crystalline solid. Any extra water more than that required to react with the cement, will remain as liquid water in the mix and will eventually evaporate leaving a porosity within the concrete. This porosity allows water and carbon dioxide to enter the concrete to cause rusting of the reinforcing steel and consequent failure of the whole structure.

A properly mixed concrete with the appropriate amount of water will be without pores, will be quite waterproof and will protect the reinforcing steel inside the structure. Concrete is naturally alkaline and that can react with carbon dioxide in the atmosphere and cause the concrete to become acidic if the concrete is porous. When reinforcing steel rusts it occupies a larger volume and this expansion causes the concrete to "spall" or break apart from the surface down. This is commonly called "concrete cancer" and is a certain result of too much water in the original concrete and also probably too little thickness of cover of concrete over the reinforcing steel.

Concrete cures (sets) very slowly and reaches suitable strength after 28 days. When concrete is poured into formwork, it may take an initial set after maybe 6 hours. Concrete for important projects such as large buildings, bridges and dams has its strength specified by the engineers who design the job. For these really important applications the concrete supplier and the concrete laying contractor have to prove to their client, the construction company or project owner that the concrete supplied was within the required specification.

In such cases an independent testing laboratory (NATA accredited) is contracted to sample and test the concrete.

The testing company takes samples of the concrete just before it is to be poured and the initial test is called the slump test which is also specified by the designers. A metal funnel is filled with wet concrete taken from the concrete agitator truck, and the funnel with the small open end at the top is lifted of the mass of concrete after about 3 seconds. As the wet concrete is in the form of a slurry it will slump when the funnel is lifted off. If excessive amount of water has been added to the wet concrete, it will slump into a pile and the height is compared to the height of the funnel. The loss in height in mm is measured.

Good concrete has a slump of about 80 mm which means that it is fluid enough to be formed in the formwork but not too wet and sloppy, which is the first indication that excessive water has been added and that its final strength would be lower than specified. After the slump has been confirmed as being suitable (or not) the testing company casts the sample of concrete into 3 steel cylinder moulds of 100 mm diameter and 200 mm high. The concrete is then usually poured into the formwork of the job if the slump test was within the allowable limits.

The concrete cylinders are extracted from the moulds, given a sample number and placed in a bath of water to complete the chemical reactions in the concrete, without any chance that the water in the concrete would evaporate, which would change the concrete properties.

The concrete compressive strength is tested by placing the moulded concrete cylinder vertically in a large press which applies hydraulic pressure slowly until the concrete fractures under the load. The maximum load just before fracture is the compressive strength and is expressed as Megapascals or MPa.

The first sample is tested for compressive strength after 7 days which gives the construction contractor the first indication that the concrete is on target for the specified strength. A rule of thumb is that after 7 days concrete will usually reach 50% of its design strength. Then the next 2 cylinders are tested after 28 days to confirm that the specified compressive strength concrete was placed in the job.

Concrete test cylinder in compressive test machine.

There are many grades of concrete for the many applications and the main difference is its compressive strength. In fact its compressive strength is the main property used in the design as concrete has a very poor tensile strength. Some main grades are shown with approximate compressive strengths:

General purpose, 20 MPa, domestic driveways and paths

Medium strength, 25-32 MPa, house foundations and slabs

High strength, 40 MPa, commercial buildings

High strength, 80 MPa, bridges and aircraft runways

Low heat, 120 MPa, dams and other very large structures

In very large structures a special slow setting, low heat, concrete is used to limit any shrinkage or expansion during curing.

Concrete test cylinder after testing. Many test cylinders split vertically and are totally destroyed.

Alan McGown

In chemical reactions a high temperature makes any reaction proceed faster and a faster reaction produces more heat and a runaway fast reaction and high temperature can result. For very large structures there is very little opportunity for heat to escape as the material is self insulating and to allow heat to generate within the concrete will result in excessive expansion and contraction forming cracks in the concrete when it is at its weakest, that is before it has developed a fully cured strength.

Some dams have water circulation passages cast into the structure to allow refrigerated water to be circulated to prevent the increase in temperature. In others the concrete is refrigerated before pouring to assure that a very slow reaction takes place.

Chapter 23
Coal mining

Key words

coal seams, bord and pillar, longwall, open cut

There are many different types of coal, but all were formed by vegetation which has become buried over millions of years, and due to compression from rocks and soil above, changes its form to the black or brown mineral we call coal. Of course it is much more complicated than that as over the millions of years, vegetation grew, died and was buried and more vegetation grew died and was buried over the top.

Coal occurs in seams which generally means a mostly horizontal layer which can vary from 20 cm to 10 metres. Within the seam there are bands or plies showing that the growing and burying process was repeated hundreds of times. In between the plies or seams of coal are bands of clay or rock.

The quality of coal is severely affected by the clay or rock called partings (horizontal bands), as this material contributes to the ash content of the coal which makes it less valuable. The best coal is called anthracite or bituminous and may have an ash content of 2% to 15%. This is generally called coking coal as it is made into coke by heating to be then used to make steel. When the coal is heated the volatile compounds escape from the coal and leave coke which is basically carbon, as described before in the section on town gas.

The coke is used to make steel in a blast furnace where it is mixed with iron ore (iron oxide) and heated to glowing heat. The reactions are varied as there are numerous types of iron oxides in iron ore and various theories about just exactly the reactions are so the following general reaction is using magnetite, Fe_3O_4.

$$Fe_3O_4 + 2C \rightarrow 3Fe + 2CO_2$$

Iron ore + carbon $\rightarrow$ iron + carbon dioxide

A large proportion of coal mined has a relatively high amount of ash which can be used in a coal fired power station to make electricity. Some power stations can tolerate an ash content of around 28 % and then as a result, have for each tonne of coal burnt, 280 kg of ash to dispose of. This is a serious problem as there are not many uses for the ash compared to the large amounts produced. Some ash also called fly ash, is used in concrete and road building, but these applications consume only a tiny proportion of that produced by the power stations. The only option is for the power stations to store the fly ash, and sometimes use local open cut mines which have been supplying coal to the power station.

Coal is mined by two basic methods, underground and open cut. There are advantages and disadvantages in each type and it mostly depends on the geology of the coal seams.

Underground mining occurs within the seam and does not allow for mining of multiple seams.

Underground coal mining

With underground coal mining there are several types of mining depending on the thickness and properties of the coal and the local geology. Bord and pillar mining is where tunnels called drives or bords are dug horizontally within the coal seam and these days the coal is transported out of the mine by conveyor belts. In the early days coal was transported in skips or trolleys on rails, either pushed by hand or pulled by horses.

These days the machine doing the tunnelling is called a continuous miner and at the front it has a large rotating drum with spikes called picks set around the circumference. This drum rotates and is pushed into the coal face which then pulls the coal down onto the front of the continuous miner. The coal is pushed onto a conveyor built onto the continuous miner and takes the coal to the rear of the machine where it falls into a shuttle car. This shuttle car is a low profile electric truck and the coal is transported to the start of the main conveyor which takes the coal to the surface.

The pillars are in fact much larger than the drives and this results in maybe only 30% of the coal being extracted as the pillars may remain. The drives are usually about 4 to 5 metres wide and pillars may be 30 metres by 30 metres. The pillars, as the name suggests, are left in place to hold up the roof or the layers of rock above the coal seam. The height of the drives is usually the thickness of the seam being mined and averages 2 to 3 metres. In some underground mines the pillars are also extracted but this is particularly dangerous as the roof then collapses. In most cases the pillars are left in place.

Continuous miner. Note the drum at the front and the seam of coal in the background

A more modern type of underground mining is called longwall mining. The mine is developed much like bord and pillar but large blocks of coal the thickness of the seam commonly about 2 to 4 metres high measuring about 1000 metres by 100 metres are left in place during the initial mine development. At one end of the block of coal a face conveyor is placed and above this runs a device called a shearer which is rotating drum with picks on its surface which dig into the coal along the 100 metre face and the coal falls onto the face conveyor to be carried to the surface.

Shuttle car discharging its load onto a conveyor.

Part of the longwall mining machinery are roof chocks. Each chock is a pair of very large steel plates measuring about 4 metres long, 2 metres wide, one on the floor and the other against the roof, which are separated by large hydraulic rams. The purpose of the chocks is to hold up the roof of the mine adjacent to the shearer and face conveyor. The chocks are lined up side to side in a line of about 100 metres and as the shearer cuts a line of coal off the face, the conveyor is advanced by the thickness of the shearer cut to get ready for the next cut. The line of chocks is also advanced one by one by reducing the pressure in the hydraulic cylinders which lowers the top plate. Then each chock one by one is moved closer to the face conveyor.

Alan McGown

Longwall mining roof support chock. The rollers for the conveyor are shown at the front of the machine.

Eventually the roof is overhanging the area behind the row of chocks and periodically it breaks off and falls, filling the space called the goaf. This is most necessary as the roof behind the row of chocks will put enormous pressure from the weight of rock above and can even force the chocks down into the floor. When the roof above the goaf collapses is makes an enormous noise and rush of air, but most importantly it reduces the pressure on the chocks and the whole area is then safer from the pressure of the rock above.

Longwall mining can extract a much higher proportion of the coal in the seam, around 70%, while bord and pillar mining averages about 30%.

Longwall coal mining machine showing the roof support chocks at the left and the shearer which cuts into the face of the coal seam. The conveyor is underneath the pile of coal under the shearer.

Underground mining is only capable of extracting one seam, while open cut mining is capable of extracting multiple seams.

Open cut or open pit mining

Open cut mining is suited to mine areas where there are multiple seams and a very high proportion of the coal in the ground can be extracted. It also allows for any seams of poor quality to be discarded.

Before any mining can commence the coal seams (both underground and open cut) are located by bore core drilling on a grid pattern where bores may be 500 metres apart for the initial series of holes. The drill brings to the surface a core of rock and the depth of the seams is easily seen and depth is measured.

Alan McGown

Drill rig and box of cores.

The coal from the cores is also sampled for testing. All of the dimensions, location of each hole, depth of the seams and coal quality are fed into a computer which then provides a graphic model of the seams. This allows the calculation of the exact location of the deposit to be shown and allows the most economic method of mining to be calculated.

The graphical model is also used when mining actually commences to where to start mining. GPS is also extensively used to keep control of the mining operation.

With open cut mining the soil and rock (called overburden) above the coal must be removed and placed somewhere before the coal can be mined. If the thickness of the overburden is too great, that will increase the cost of the mining. The overburden is removed by either large draglines (earth moving machines) which dump the overburden back into a previously mined area or by a large hydraulic excavator which loads large mining trucks to carry the overburden away. There are advantages and disadvantages of each method, depending on the geology of the overburden and the coal seams.

There is a lot of work in removing the overburden, and the thickness of the coal seam and the coal quality must justify the expense.

Open cut brown coal mine.

Once the seams of coal of interest are exposed, usually an excavator an digs into the face of the coal seam and mining trucks carry the coal to a crusher, to reduce the coal size to a maximum of 63 mm. Coal in the run of mine (ROM) stockpile adjacent to the crusher may have large lumps over 500 mm which is too large for conveyors to handle without damaging the conveyor belt.

Coal washing

Many mines are located near power stations and if the quality of the coal is suitable it can be supplied directly to the power station in the form that it comes out of the mine. But in many cases the ash content of the run of mine (ROM) coal as it comes from the ground is higher than the specification required by the power station. A typical specification for ash content may be 24% by weight. If coal as mined contains large amounts of rock and clay, its ash content may be above that limit and that coal may have to be discarded. Some mines are able to blend high ash coal with lower ash coal to reduce the average ash content to below the allowable limit in the specification. That is not always possible and to reduce the ash content the coal needs to be treated in the washery, called in the industry the Coal Handling and Preparation Plant (CHPP).

The first step in the CHPP is that the coal is crushed usually to a size below 63 mm. At this stage there are lumps of coal close to 63 mm and all the way down to a fine powder. There are several methods of separating the high from the low ash coal depending on the particle size.

The coarse coal is mixed with water and a mineral called magnetite, which is a form of magnetic iron oxide. Without the coal this mixture of water and magnetite takes on the properties of a dense liquid and this density can be adjusted to whatever level the CHPP operators desire, typically 1.6 g/cc. Any coal of density below 1.6 (low ash) will float and higher density particles will sink. The mixture is passed as a slurry through a hydrocyclone which separates the rocks from the coal due to the large density difference. Coal has a relative density of around 1.3 up to 1.8 g/cc while rock has a density above 2.3 g/cc. This separation sounds easy, but in fact there is a continuum of coal containing varying amounts of ash. The higher the ash content the higher is the density and the CHPP operators must make the decision as to what density and thus what ash content will be accepted and what will be rejected. Typically the reject limit is set at 1.6. The low density stream flowing from the hydrocyclones is sent to the stockpile and the higher density stream with high ash content is sent to the rejects pile.

Usually the last step is in the froth flotation device which usually handles the finest particles of coal. Diesel oil and a detergent is mixed with the fine coal suspended in water and air is blown into the bottom of the tank. Coal being hydrophobic adheres to the diesel and then the bubbles carry it to the surface as a froth to be skimmed of as good quality low ash coal. Any high ash material does not stick to the diesel and air bubbles and sinks to the bottom where it is separated and sent to the rejects pile.

So the term coal washery is a misnomer as it is simply separating low density (low ash, high quality) coal from the high density (high ash, low quality) coal and rocks.

Alan McGown

Chapter 24
Sampling and testing

Key words

sampling, representative sample, homogeneous,
calorific value, calorimeter

This is a very important issue when testing products,
as the test must be done on a sample which is
representative of the whole

Another point is that testing must be done in a timely
manner so that the result can be used to make
adjustments and possibly save the life of a person in
hospital.

When a nurse takes a sample of a person's blood for
testing, it is assumed that the whole of the blood in the
person's body is homogeneous, i.e. it is all the same,
and any sample has the same properties as the rest of
the blood in the person's body. It is not as easy in
industry as the product to be tested may vary widely.

This chapter describes the sampling and testing of
coal, but the principles used in sampling are applicable
to many industries where the products being tested are
not homogeneous, that is one part of the product may
be different to another part. There are many examples
where this is the case – minerals, grains,
manufactured products, oils, bitumen, road making
gravel, town gas etc.

Consider a stockpile of coal of say 5000 tonnes of coal which is going to be delivered to a power station to burn to generate electricity. One lump of coal is always different to the next lump, that is the material is not homogeneous and even one shovel full will be different to the next shovel full. So a way needs to found which can guarantee what is tested really represents the whole batch.

The price of the coal to be paid by the power station to the mining company depends on many things, but particularly the ash content.

Coal for power stations may contain 15 to 30 % ash, which means that at say ash content of 30% there is only 700 kg of coal in each tonne delivered, so a lower price would be paid for the coal of 30% ash than for coal of 20% ash.

How can it be determined what is the ash content of the 5000 tones being delivered on one day to the nearby power station? Coal in the ground as mentioned before, consists of plies or layers within seams and these plies and seams may be separated by bands of clay or rock (called partings) which are about 100% ash.

When the coal is mined there is an unavoidable amount of partings (rock and clay between the plies) which finds its way into the stockpile, as the excavator driver cannot avoid also digging up some of this partings material with the coal. The good coal of low ash and rocks of high ash are all mixed up together in the stockpile. If the coal is passed through a washery as described above then the average ash content is reduced. But we still need to know the measured ash content of the stockpile so the right price can be charged by the coal mine.

When the coal arrives at the power station by truck, train or conveyor belt, it is dumped into the receiver hopper and then passes along a conveyor belt equipped with an automatic sampler. As the coal is flowing along the conveyor belt, a small portion say 1 kg is diverted by the sampler say every 3 minutes (20 samples per hour) into a sampling bucket. If the 5000 tonnes took 8 hours to be delivered then there would be 160 kg of coal in the sample bucket. Most automatic samples have a secondary crusher/ sampler and that 160 kg would be reduced to say 16kg.

That 16 kg of coal sample now truly represents the 5000 tonnes of coal and any tests done on it will be applied to the 5000 tonnes. That means the average composition of the 16 kg of coal is the same as the average composition of the 5000 tonne batch.

But when the laboratory tests the coal it needs to weigh out test samples of about 1 gram, and as the coal contains also pieces of rock. One cannot simply take out with a spoon those 1 gram subsamples. In addition, coal samples will have lumps up to 50 mm.

The laboratory then passes the whole 16 kg sample through a crusher to reduce the 50 mm lumps down to 4 mm lumps. Then the 16 kg of coal is passed through a rotary sample divider which reduces the 16 kg of 4 mm sized coal to about 300 grams. Now each 4 mm lump may still be good coal, high ash coal or rocks. Still too large to weigh out 1 gram samples for testing.

The 300 grams of 4 mm coal is now pulverised to a fine powder and after mixing, each 1 gram sub sample will have exactly the same composition as any other 1 gram sample. Now the testing for ash and other properties can begin.

The most common tests on coal are ash content, moisture content, sulphur content and calorific value. There are many other tests used in the industry, but these described are the most often used.

All tests carried out are usually described by an Australian Standard method.

Ash test

An approximately 1 gram of pulverised coal is accurately weighed into a ceramic crucible and then this crucible is placed into a furnace at about 800 degrees Celsius. The volatile compounds in the coal are evaporated and pass out of the furnace and then the carbon in the coal reacts with oxygen in the air until only the ash is left in the crucible. After cooling, the crucible is weighed again and the ash content as a percentage is calculated by dividing the mass of the ash by the mass of coal in the test.

The ash is usually a mixture of silicon dioxide, iron oxide and aluminium oxide with a long list of other minor metal oxides.

Moisture content test

Similar to the ash test a small sample is weighed into a small dish which is placed into and oven at 105 degrees, which causes any moisture to evaporate.

After cooling, the dish containing the dry coal is weighed again, and the moisture content is the difference in weight compared to the weight before heating.

Sulphur content test

Again a small sample is weighed into a small ceramic boat. This is placed into a furnace at around 1000 degrees in a flow of pure oxygen. The gas generated which is mostly carbon dioxide and sulphur dioxide passes into a glass cell through which are being passed infra red radiation. The amount of radiation absorbed is proportional to the percentage of sulphur dioxide and thus the sulphur content of the coal can be calculated.

Volatiles test

Again just like the ash and moisture tests the volatiles test requires a small sample to be weighed and placed in a furnace at 900 degrees, but in this case the sample is in a closed crucible which will allow the volatile contents to pass out through the lid, prevent oxygen from entering and leave just carbon inside the crucible. The amount of lost weight is called the volatiles content.

This test is particularly important for coking coal which is going to be turned into coke for steel making.

Calorific value test

This was described in chapter 3.

Sampling is also very important in most other industries where tests are to be done on a sample to determine its properties. Just like sampling coal, any sample to be tested must truly represent of the batch of material it comes from. The term "representative sample" is very often mentioned. If the operator doing the sampling cannot guarantee that the samples tested are representative then there is no point in doing the tests.

All testing is done for quality control purposes, and if the material being tested is outside of the specification, then those people controlling the process need to know, so that they can modify the process to bring the properties back to within the specified limits.

A major part of a NATA accreditation as described in chapters 21 and 22 is the need to demonstrate adequate sampling procedures.

Conclusions

When you study any branch of science, it will not give you the exact information you may need for any particular situation, but it will give you the ability to find and learn what you need to know for any particular situation.

My experience, in addition to general laboratory testing in various industries, has been in carrying out factory trials of industrial chemicals at customers factories, and in these situations many unexpected things happen. When something does not happen according to plan, then you need to be able to figure out what is going wrong, find possible causes and remedies

It is only by application of scientific thought and examination of the facts, that the cause and a remedy may be found. *"The scientific method"* allows one to make deductions and predictions based on your observations. Then you need to see which of the possible causes fits correctly with your observations.

As I mentioned in the introduction, there is science in everything around us, and a good understanding of Science along with Maths and English will greatly enhance your employment opportunities.

Author profile

Alan McGown is an industrial chemist who has worked in all of the industries described in this book. He started as a laboratory assistant, graduating to laboratory management, then on to technical marketing management conducting trials of industrial products and chemicals in customer's factories, and finally returning to working at the bench setting up and operating three new laboratories for testing coal.